Boo
ab

CLEVER DOG

CLEVER DOG

The Secrets Your Dog Wants You To Know

Sarah Whitehead

Collins

First published in 2012 by Collins
HarperCollins*Publishers*
77–85 Fulham Palace Road
London W6 8JB

www.harpercollins.co.uk

1 3 5 7 9 10 8 6 4 2

A catalogue record for this book is
available from the British Library

ISBN 978-0-00-744408-3

Printed and bound in Great Britain by
Clays Ltd, St Ives plc.

MIX
Paper from
responsible sources
FSC C007454

FSC™ is a non-profit international organisation established to promote
the responsible management of the world's forests. Products carrying the
FSC label are independently certified to assure consumers that they come
from forests that are managed to meet the social, economic and
ecological needs of present and future generations,
and other controlled sources.

Find out more about HarperCollins and the environment at
www.harpercollins.co.uk/green

DISCLAIMER

In the interests of protecting client confidentiality, the names of people,
places and dogs have all been changed – apart from mine!

DO TRY THIS AT HOME!

All of Sarah's methods, if carried out appropriately, are kind and safe for dogs
of all ages and types. However, if your dog has an aggression problem or his
behaviour is impacting on his – or your – welfare, then professional help is needed.
Ask your vet for a referral to a behaviour specialist in your area.

To Tao; my heartbeat at my feet

Acknowledgements

To my wonderful agent, Philippa Pride, for her dedication and terrier-like determination, and her Labrador, Elvie, for his contagious good humour.

To John Fisher – sadly missed – who still inspires me to 'Think Dog' every day.

To commissioning editor Julia Koppitz and all the team at HarperCollins for sharing my vision.

And finally, to all the dogs and owners who have taught me so much, and continue to do so.

Contents

Part 3: HEALTH

Part 4: HAPPINESS

Introduction

Dogs and dinner parties

Dogs are a household subject and one – it seems – on which everyone has a fascinating opinion. Numerous dinner parties attest to this. There I am enjoying a nice glass of Chablis and a friendly chat when someone asks what I do for a living. I've never yet said 'I'm in IT' but I may well consider it in future because no sooner have I confessed that I'm a pet behaviour specialist than someone comments: 'Ooh, that's great. I have a dog you could help with.'

At this stage, of course, they don't really want my help – they just want to talk about Bonzo's latest naughty adventure and, frankly, I love a good dog story. Once they are done, the comments from the other guests are revealing. Having heard how Bonzo bites people, runs off, barks all day or eats the carpet, someone will always say, 'I know how to stop him doing that' – and that someone is never me! Invariably, the 'expert' will proclaim the most hideous, punitive measure to be a complete and universal cure. Apparently, it's OK to suggest to a complete stranger that they use an electric shock collar on their dog, that they should rub their dog's nose in faecal matter or that they should strangle the poor thing on a choke chain, all in the name of training. Uh oh.

At this point in the conversation, there's a small pause. The circle of guests stop and turn to me. 'So, what do you think of doing that?'

Pass another glass of wine, please, and make it a big one.

Strong opinions on dog behaviour are not confined to the social scene. Over the years, I've appeared as an expert witness in court cases where the magistrates have, quite literally, the power of life or death over someone else's pet. These cases are often depressingly complex, costly and time-consuming, and yet, all too often, after I've given evidence on a specific behavioural aspect, the magistrate will say something along the lines of: 'Thank you, Ms Whitehead, but I think we know how dogs behave.'

My work as a pet behaviour counsellor has kept me busy, impassioned, fascinated and laughing for the past twenty years. In that time, I have had the honour of meeting thousands of dogs and owners across the UK and internationally, and have relished the chance to 'talk dog' with them. I now specialise in aggression problems and the more weird and wonderful behaviours dogs can present, while my practice deals with every kind of canine behaviour problem that you can think of – and then some. Working on referral from veterinary practices around the country, we are constantly in touch with the practicalities of canine training and behaviour – and still, after all these years, I love every minute of it.

Perhaps the fact that dogs are so common in our culture works against them. They pervade our domestic existence and yet they are remarkably under-researched. Want to find out about dog behaviour? Nine times out of ten you will be forced to read literature on wolves, studies on rats and research on pigeons. This is the equivalent of studying human psychology by examining the behaviour patterns of the great apes. It may be interesting and relevant, but it isn't the same as studying the actual species that we live with day in, day out.

Maybe the scarcity of rigorous scientific research on domestic dogs is the reason why so many myths about their behaviour pervade our society. In many cases, these myths have become so ingrained that we accept them as truth, never questioning where they came from or how factually correct they actually are. In this book, just as in my everyday work as a pet behaviour specialist, I will propose some new ways of thinking about dogs. Some of these may be thought-provoking, some even shocking. In an age when we are all too often blinded by the ideas we are fed by television, the lures of urban myth and the promise of a fifteen-minute fix, it's all too easy to think about dogs from a purely human perspective. Instead, perhaps now is the time to think about life from the dog's point of view: to let go of all the previous theories that you have toyed with, to open your mind and, as my mentor the late John Fisher always said, 'Think Dog'.

Part 1

Life

1

Dominance

The wolf legend and other myths

Case history: **Ice, the misunderstood Malamute**

'This is a wolf you are dealing with here!' barked the training instructor. 'You need to show him who's boss.' With that, the big man strode across the room, grabbed the lead from the humiliated owner who had been struggling to get her dog to sit, and strung him up so that his front feet were off the floor.

At eight months old, Ice, the big, grey-and-white Malamute, had never been treated like this before and did what any reasonable dog would do when it thought its life was under threat: he became very still, averted his gaze, and uttered a long, low warning growl from deep in his throat.

'No dog's gonna growl at me,' stormed the instructor, and with one swift movement he launched himself at the dog and managed to wrestle him to the ground and onto his back. Ice wet himself in fear.

'There,' said the trainer. 'He's submitted.' He got up and wiped the urine from his sleeve.

Ice stayed where he was, lips drawn back and tail curled defensively under his belly.

'Now, you need to do that every day,' said the instructor. 'It's called an alpha rollover. It's what the pack leader would do to the other wolves in the pack to keep them submissive. You need to act like a pack leader, and doing this exercise every day is part of it. OK?'

Ice's owners, Keith and Sharon, nodded in quiet agreement. They took back the lead and sat down at the side of the hall with a very subdued dog by their side. It wasn't nice to see their pet being manhandled, but it did seem to have an effect on him. After all, he was a big, powerful dog – Malamutes may look like Huskies, but they are bigger and stronger – and he had started to pull them towards strangers in the street. They knew that they couldn't allow him to become out of control.

Keith and Sharon did their homework every day as instructed – or at least they tried to. Over the following week, they subjected Ice to repeated 'alpha rollovers'. The first day, Ice seemed to think it was a bit of a joke. He took Sharon's arm in his mouth and held it as if he was playing, but she persisted and forced him onto his back. He lay there looking bemused, and jumped up again as soon as she let go. On the second day, things weren't so easy. As Sharon approached, Ice dodged out of the way and tried to flee the room. Between them, Keith and Sharon caught the huge dog and pinned him down, but he growled continuously and Sharon was sure he tried to snap as they let him get up. By the Thursday – you've guessed it – Ice was having none of it. He had tried every trick in the doggie book to get his owners to stop behaving so weirdly and his patience had run out. When they approached him, Ice bared his teeth and snarled so ferociously that Sharon was genuinely frightened; Keith probably was as well – and who could blame him? They decided to leave Ice alone.

The next night, they were back at the dog training class. For the first time since joining the class a few weeks before, Ice stiffened as he entered the hall and growled at a lady who reached out her hand to stroke him. This took Keith and Sharon completely by surprise, as he had always been friendly with the other owners before. The fray attracted the attention of the instructor who came stomping towards them, red in the face and looking as though he meant business. This time, however, Ice was ready for him. As the instructor leant forward to take the lead from Keith, Ice saw his opportunity and lunged at him, teeth on full display and barking (well, spitting) with full force. Unfortunately, at this point the instructor made a terrible mistake. Instead of backing off, turning away or reducing the level of threat he was showing the dog, he moved towards him, determined in his own uniquely human way to have the last word. His lesson was a clear one. Ice bit him on the arm. With Sharon in tears, Keith shaking with adrenaline and Ice in disgrace, they were ordered to leave the hall, the instructor's final words ringing in their ears: 'That dog is out of control. He's dominant and aggressive and you should have him put down. If you don't, I'll report you.'

Keith and Sharon sat on the sofa in their front room as they relayed this sorry tale to me. Pale and anxious, they frequently glanced at each other for reassurance, clearly worried about what I was going to say. They feared they were going to lose their precious dog.

Sadly, this is a story I am all too familiar with, in one form or another.

It's so tempting to believe the traditional story: a caveman tames a wild wolf and they both live happily ever after. However, this myth fails on so many counts that it's a wonder it has prevailed for so long. Domestic dogs are not simply tame wolves. Nor do tame wolves ever become 'domestic', no matter how well they have been hand-reared. Although dogs and wolves share a genetic background, failing to distinguish between the two species is as misguided as failing to distinguish between man and ape.

So, how did such myths come to be so prevalent in our society? Why is it that so many books, television programmes and self-styled 'experts' are still claiming that dogs are really the 'wolf on your sofa'?

Many of the tales told about wolf behaviour are simply untrue. The idea that the alpha pair always eat first is one example. If this were truly the case, how on earth would wolf cubs survive? Such a claim simply doesn't make evolutionary sense. The fact is that wolves are highly social and a kill is shared amongst the pack – with youngsters, adolescents and even elderly wolves getting a good meal, often before the other adult members of the group.

Other theories – such as the idea of a linear hierarchy – where a strict ladder of rank exists – are based on captive packs. However, keeping any kind of predatory animal in a restricted area with others of its own kind is a far cry from 'natural' behaviour; we only have to watch the *Big Brother* television show to appreciate that!

Even the idea of submission has been woefully misinterpreted. Wolves living a free and wild existence don't exert dominance over each other in order to elicit submission; on the contrary, submission is offered freely by one wolf to another in order to gain reassurance – something quite different – and it's a behaviour that's not shared by their domestic cousins, the dog. Indeed, wolf cubs raised with domestic dogs sometimes illustrate this in the oddest ways, with the cubs trying to push their muzzles inside the jaws of their rather bemused domestic dog elders.

While some of the misinformation that exists about 'wolves versus dogs' is simply amusing, in my role as a canine behaviour specialist, I have first-hand evidence of the problems such ingrained and unquestioned beliefs can cause. The idea that dogs view humans as a part of a pack, and that they must observe pack rules in order to be 'leader', has led to some extraordinary urban myths. These include encouraging owners to sit in their dogs' beds to show them who's boss, and telling them to ignore their dogs when they first come home in case the dog tries to 'dominate them'. Owners are asked to pretend to eat before feeding their dog in a misplaced effort to show the dog he's bottom of the pile. Some even follow the advice to pin their dogs down, in a so-called 'alpha rollover'. What dogs think of these appalling misjudgements of their communication systems is anyone's guess. How they put up with these bizarre human ways is even more of a mystery, but to me, the greatest travesty is that very few people stop to question the rationale behind the rules they are given.

So, if dogs are not wolves in disguise, what are they? Well, there is strong evidence to suggest that dogs are just big babies! Instead of developing into full-blown adult wolves, evolution has caused a genetic shift, bringing about both physical and behavioural changes. Domestic dogs' skulls are smaller than wolves', their teeth are relatively undeveloped and their reproductive cycles are different. Domestic dogs yelp, they whine and they bark – all characteristics that are lost by the time a wolf is five months old. In other words, dogs stay for ever as juveniles – playful, puppy-like and highly dependent on their parents.

Dogs and humans share a long and successful history together because of the ability to get along, and because they need us. This sociability is based on

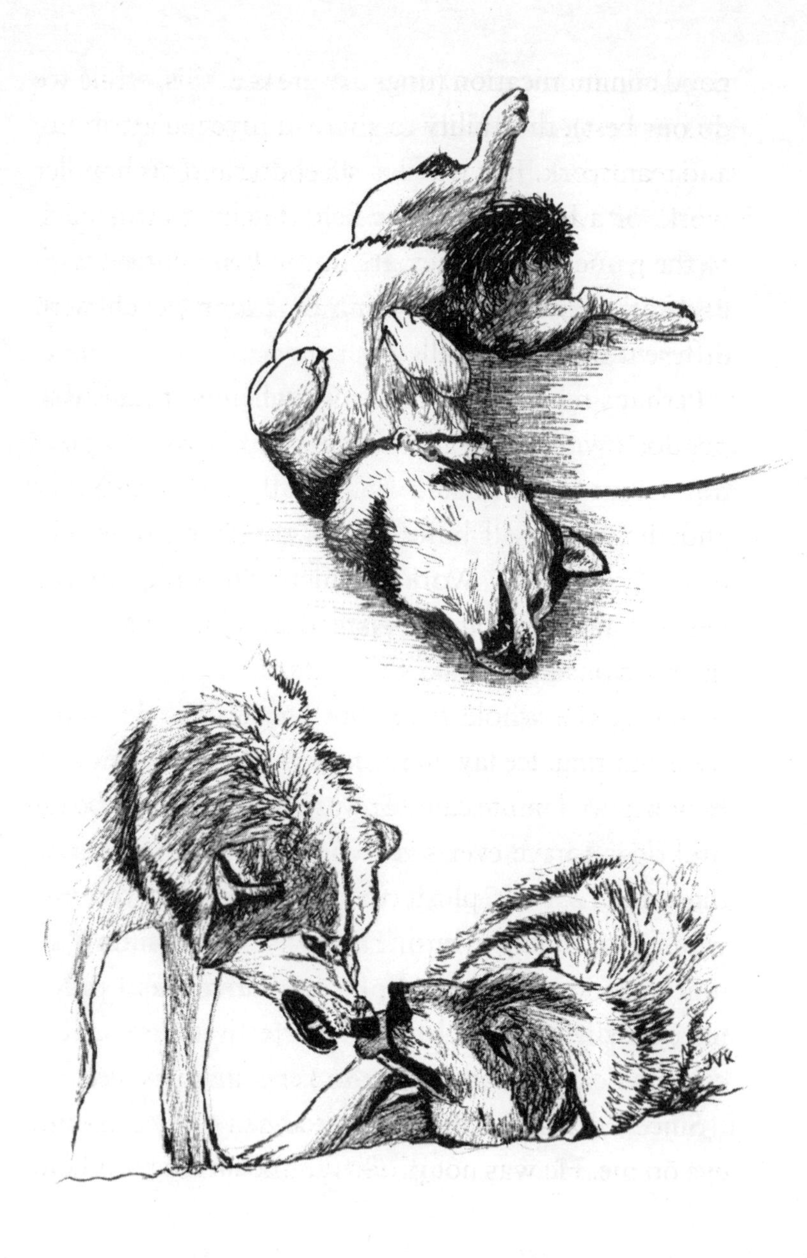
JVK
JVK

good communication (dogs are great at this, while we do our best), the ability to share (a juvenile attribute) and teamwork. Just watch a sheepdog and his handler work, or a Retriever in the field, fetching birds back to the gamekeeper. They are not in competition with each other but are operating as a team, each with different but equal skills.

Perhaps my greatest challenge when working with pet dog owners is to get the message across that your dog is not an adversary but an ally. Work together and the team will build bonds so strong they will never be broken. Work against each other in the belief that your dog is trying to dominate you, and the relationship will start to suffer.

During the whole time that Sharon, Kevin and I were chatting, Ice lay on the floor looking nonchalant, as only a Malamute can. He was one of the most beautiful dogs I have ever seen, with ice-blue eyes (hence the name), a thick, plush coat and a face that you just wanted to snuggle into. However, every time I so much as moved my hands, he became still and rolled his eyes slightly – a clear and perfectly polite canine communication warning me to keep my distance.

Since I'd entered the house, Ice had kept a careful eye on me. He was not proactive and as long as I held back, I knew I was going to be safe around him. He just needed to know the same about me. It was little surprise that Ice thought I was likely to be bad news. Ever since the training class incident, Keith and Sharon had received few visitors, and had taken to walking Ice in the early hours of the morning so as to avoid other people, fearful that he would act aggressively and force them into making a final decision. Poor Ice had been kept a virtual prisoner in the home, and although he was well cared for, his owners had been 'tip-toeing' around him, fearful of what he might do next.

As in many such cases, the road to success would require a careful combination of planning and action. Ice desperately needed an active social life in order to re-establish his social skills with other people and other dogs. But in the early stages, my uppermost goal was to build the family's confidence in each other once again. We started with some basic training. Training is often under-rated in these situations, but it can be the glue that holds the family together while relationships are reformed. Thankfully, training these days can be fast, fun and friendly. Gone are the days of 'stomp and jerk' techniques. Indeed, what we know now is how much faster dogs learn if they are encouraged to use their brains rather than brawn. Nowadays, there are many enlightened trainers and instructors out there who are using behavioural understanding to underpin their training – a far cry from the outdated techniques that poor Ice had been subjected to.

Ice was a fast learner. With the help of some tasty morsels of cheese, we had him sitting up and (quite literally) eating out of my hand within minutes. Having motivated him to want to engage in some fun interactions with his owners, we encouraged him to sit, lie down and give a paw on command. We finished with a finale of 'be a bear' – a great trick in which the dog sits up in a begging position and looks as cute as cute can be.

Inspired, his owners were given two weeks to practise these new behaviours. Ice was to be engaged in 'fun training' twice a day, using food and a relaxed approach, and was to get used to wearing a special head collar so that his owners could take him for a walk without being towed along behind him. Head collars allow almost complete control, like power steering, and make perfect sense in cases where the owner needs to avoid getting into a battle of strength with a big dog, who is clearly going to win. After all, no one would consider walking a horse on a piece of string. This bit of kit made all the difference to Sharon, as she could walk Ice with confidence when using it.

After two weeks of indoor practice, we were ready to hit the streets – and it didn't have to be at two o'clock in the morning either. Poor Keith and Sharon had lost so much confidence in their ability to handle their dog that we formulated a programme of 'stop on sight'. This meant that every time we saw some-one walking our way, we asked Ice to give us atten-

tion, stop and sit. How, you might ask? Through good, old-fashioned bribery. In fact, this is not the 'giving in' that many people think it is. In the first few instances when Ice saw someone heading towards him, we simply showed him that we had some delicious chunks of cooked chicken on offer. This immediately produced the desired result, However, the next part was pure magic.

Imagine you have been summoned to see the boss. Generally, when this happens it's not good news. As you walk towards his or her office door, you start to feel a little anxious, your palms get sweaty and you begin formulating defensive arguments in your head to stave off possible attack. However, when you go in, your boss greets you with a big smile and holds out his hand to congratulate you. You have earned a bonus. He gives you £1000 in cash on the spot. You are overjoyed! On the same day the following week, you are once again called to the boss's office – and the same thing happens all over again. Unbelievably, this is repeated a third time, at the same time and on the same day of the week. Now, just imagine how you feel going into work in the fourth week. Are you elated? Full of hopeful expectation? You bet! Rewards that are linked to circumstances (days of the week, visual cues, people) have the power to affect your emotional state, not just your behaviour.

With this in mind, we worked on the equivalent canine scenario, and watched as Ice made giant conceptual leaps. During the following six weeks, the appearance of someone walking towards him in the street prompted him to whip round, sit down and look up at his owner, without having to be bribed, cajoled or reminded. Even better, he even started to offer 'be a bear', which elicited smiles and laughter from those walking past where they might once have given him a wide berth. These new responses were heavily rewarded. We played with Ice in the park, the street and in the communal front garden of his owners' home. We gave him chicken, dried liver, hot dog sausages and fish treats for being calm and social around new people both inside and outside his home. We praised him for good behaviour and ignored it when he got it wrong. It worked like a dream. Within three months, without any drama, pretending to be Alpha leader, punishment or Alpha rollovers, Keith and Sharon found themselves back in control – and completely besotted with their lovely dog again.

TOP TIPS ON BUILDING A GREAT RELATIONSHIP WITH YOUR DOG

» Work on building a trusting relationship with your dog. Don't assume he is trying to challenge you for 'leadership'. The best dog/owner combinations are teams, not competitors.

» Start as you mean to go on. If you don't want your huge, muddy dog getting on the sofa when he's three years old, don't let him do it when he's twelve weeks old – no matter how cute.

» Start training early, the second that you can – especially with big breeds. The old adage that you can't start to train a dog until he's six months old is wrong. Just think how much easier it is for children to learn a new language than it is for adults.

» Be consistent. Agree rules and boundaries within your family – and stick to them. Write them down if it will help to avoid domestic arguments later. Dogs like to know exactly what they can and can't do.

» Choose a training class that uses kind, fair and effective methods. The days of choke chains and 'yank and jerk' training are long gone.

» Use brain, not brawn. If your dog tries to manipulate a situation by engaging in a battle of strength, immediately disengage and use your superior intellect to defuse the situation. Many dogs enjoy physical confrontation – so you will 'win' by refusing to compete in this way.

» Replace negative commands with positive ones. For example, ask your dog to sit rather than nagging him not to jump up.

» Clicker training is a great way to teach dogs new tricks that can have an effect on their emotional states and the way they behave generally, not just their immediate actions. They also impress humans.

....

Clicker training is fast, fun and kind, and can be used in all sorts of ways. The clicker – a small plastic tool that makes a double click sound when pressed, effectively acts as an 'interpreter' between human and animal, marking the behaviour that earned the reward, and making the whole learning experience one that is focused on trial and success, rather than trial and error. This, of course, has an impact on emotion, too. I know how my dogs look and behave when I get out the clicker to do some training – it's the highlight of their day, and it also has the effect of making me feel happy, too. Want to try clicker training but don't know where to start? Find a good instructor or class at www.apdt.co.uk or watch easy-to-follow demos on the Internet. My favourites are

www.trainyourdogonline.com and www.clickertraining.com, where Karen Pryor demonstrates by training a fish. Try it for fun – you'll be amazed what your dog can do.

Over the past ten years, there has been a huge increase in awareness about animal behaviour – and, in particular, dog behaviour. Not a day goes by when there isn't some news item, television show or a headline that involves dogs – and this has made my life as a behaviour specialist both easier and more difficult at the same time. On the plus side, owners now know that they can get help for their pets when things aren't going according to plan. Those TV shows make my phone ring! On the minus side, the idea still endures that dogs are attempting to take control and are secretly trying to find ways to dominate their owners – despite the fact that their lack of opposable thumbs makes it impossible for them to open their own dog food!

There's a simple but extremely effective way of keeping an open mind when watching dog programmes on TV: watch the dogs, not the people. Dogs don't lie. They don't try to look good for the camera. They don't nod and smile when they feel dubious and uncomfortable. Dogs express their emotions in ways that are unique to canine society, but which are also remarkably similar to those that we know and understand. With a little extra effort, humans can become proficient in speaking 'canine' and they can then have free-flowing and honest communication with their dogs. Next time you watch a TV show about dogs, look at an internet video clip or gaze at your own dog, try to decide what he might be saying and what message he is trying to get across. I bet you any money it won't be a statement about rank, challenge or dominance; it's more likely to be a plea for understanding and the desire to be a player in a well-balanced team.

2

Canine communication systems

Learning the lingo

Case history: **Dave, the fearful German Shepherd**

Dogs talk. There's little doubt about that. However, on the whole, humans are bad at listening. From the owner who complains that their dog looks guilty when they come home to find their sofa chewed, to the person who is shocked that their dog was wagging its tail while barking aggressively at a visitor, miscommunication is rife.

In order to live with us peacefully, dogs have to learn about the oddities of human behaviour: the fact that we wave our arms about when we speak, that we like to greet each other by hugging (a sexual or threatening gesture to a dog) and that we disregard olfactory communication almost entirely (probably best).

However, in return, not many humans bother to learn 'canine' as a foreign language. This is a shame, because dogs communicate in ways that we can hardly conceive. For example, they can register huge amounts of information from scent – gleaning messages about another dog's sexual status and health, and how long ago they were in town – in much the same way as we get information from reading a newspaper. While this may be out of our range, watching dogs and reading them is not – and it's something that's addictive when you know what you are looking for.

Just like us, dogs can use both long-distance and intimate communication to express themselves. While we might plaster words on an advertising billboard or make a phone call to get our message across great distances, dogs use scent and big visual signals to communicate over time and space. Ever seen a dog scratch up the area where he has just urinated or defecated? He's leaving a clear visual marker to emphasise his olfactory point. Ever listened to a dog barking or howling and heard another one reply? He's just made a long-distance call.

In common with humans, dogs use subtle facial expressions and body language to communicate close up. Some expressions are remarkably similar to ours, and need little or no interpretation. Despite huge variation in the physical appearance of dogs, no matter what their sizes and shapes, we can all recognise a dog that looks miserable, is sad or in pain. Equally, we can assume that a dog with his ears up, his face rounded and relaxed, and his mouth pulled back into a 'grin' is showing happiness – just as we do (although most of us manage without moveable ears).

Some of the other signals that dogs use to communicate may not be quite so obvious to us. A micro-expression is a brief, involuntary facial expression or body movement, which reveals the emotional state hidden inside. In humans, this is most noticeable when someone is trying to conceal how they are feeling and a tiny, almost imperceptible change of expression gives it away. In dogs, there is no attempt at concealment but they can act and move much faster than us so the change can be a subtle micro-expression: a fleeting second of stillness, a tiny turn of the ear, a slight widening of the gaze or the closing of a muzzle can all give us clues as to how they are feeling.

Dave was brought to me by his worried owner. At the age of two, he had been re-homed from the local rescue centre where he had been living for some fifteen months. A handsome dog, Dave had been out to a new home and then back to the rescue centre several times. Panting heavily, he sat in my office like a dog on hot coals, watching every move I made. His face was a picture of stress and anxiety, his skinny body hunched and rounded, his tail tucked under. I hate seeing dogs like this. It's a miserable state to be in, and although we might have to accept it once in a while when the dog is ill or has to undergo veterinary treatment, there is something deeply disturbing about seeing a dog so unhappy most of the time.

George had now owned Dave for nearly five months. He had kept German Shepherds in the past, and lost his heart to this big dog he had found in the rescue centre where he was a volunteer dog walker. Dave had come into the kennel environment as a nervous but testosterone-charged adolescent. He was big and out of control, with no manners and no training. He jumped at everybody, and wrenched their shoulders out of their sockets when they tried to walk him. Worse, he began to hover in the back of the kennel then launch himself forwards with a volley of barking – enough to send all but the most experienced staff scurrying for safety. He hated being handled, and would put his jaws on anyone who tried. His future didn't look good.

Like any other animal, dogs are programmed to ensure their own survival. When faced with an immediate threat, the brain doesn't take the time to ponder carefully all the possible outcomes of the situation but simply kicks into survival mode and causes the animal to react. This state is one we can all relate to. Nearly all of us have had an incident in our lives when we were scared and simply reacted out of self-preservation. Perhaps you have had a near-miss (or should that be near-hit?) car accident. Perhaps some-one has threatened you, or you have been frightened by an animal (albeit a spider in the bath). On these occasions, the basic, most primitive part of your brain – known as the amygdala – simply takes over in order to keep you safe. Instead of information being trans-mitted to your cortex – the thinking, cognitive bits of your brain that allow you to do Sudoku puzzles and decide what to have for lunch – a message goes directly to your amygdala in what is described as the 'fast and dirty' route to reaction. It is your amygdala that makes you leap out of the way of a falling tree branch, swerve to avoid that oncoming car, and swiftly jerk your hand away from the bath plug when you notice the spider.

The way an animal acts when under pressure will depend on a number of different factors: the species, the type or breed, the individual temperament or personality of the animal, its previous history and emotional state, and the circumstances of the actual event. The coping strategy an animal uses in a moment of crisis is one of what we call the 'four Fs'. We have all heard of 'flight' and 'fight' – the strategies of running away or engaging in conflict – but there are two more Fs: 'freeze' and 'flirt'. Freeze is the most common reaction to threat but as it is often fleeting and humans are poor at noticing it, it is frequently overlooked. Flirt is commonly seen too. Just think about a dog that would prefer you didn't look in its ears; instead of putting up a fight or running off, it runs around manically, brings you a toy or leaps about like a puppy.

Your amygdala is your greatest friend. It keeps you safe and sometimes saves your life. Unfortunately, it can also become your worst enemy. Watching a scary horror movie might make you jump during the film, but afterwards – when your amygdala is still wired – it can make you leap out of your skin at the imagined sound of a footstep on the stairs. Over-activating the brain and body's flight or fight responses leads to over-activation of the brain's recognition of threat signals, and if this situation becomes chronic, it can lead to weight loss, immune deficiencies, stomach disorders and skin problems.

While in kennels, there's no doubt that Dave was anxious and fearful. Faced with the choice of running

away or standing to fight, I'm sure he would have preferred to high-tail it into the distance but the confines of the kennel walls prevented this and he began to oscillate between backing away and lunging forwards. Sadly, this is often how dogs learn to use defensive aggression and once they discover that it makes the 'threat' go away – exactly the impact they want – it quickly becomes an ingrained behaviour pattern that can be hard to break.

To most humans, barking is threatening, annoying and largely meaningless. It's something we want to stop rather than listen to. Ironically this is probably what most dogs would say about human speech! However, tune in to barking and it's possible to hear that the intonation varies between barks and that even the frequency and rapidity of the sound communicates the dog's emotional state and something of his message. A single, low 'ouff' is a quiet warning that tells of danger or potential threat. Known as an alarm bark, it is designed to warn others in the social group without giving away the dog's position. Rapid-fire, joined-up barks tend to be defensive: they say 'keep away'. High-pitched staccato yaps are more likely to be playful and show excitement. Such interpretations may seem subjective, but research has shown that even people who don't own dogs can tell the difference between a dog barking because it is joyful and one which is upset or lonely, just by listening to an audio recording.

Poor Dave was in survival mode from morning until night. George told me the dog even seemed to sleep with one eye open, and he was clearly suffering as a result. He was thin, despite being well fed. His coat condition was poor, with his hair coming out in handfuls – yet another sign of stress. He had repeated bouts of stomach trouble, for which the vets could find no clinical reason. It wouldn't be long before his chronic stress began to cause a serious degeneration in his health.

In order to work with Dave and George when they came to my clinic, I had to find a way to move about safely in his presence. Although he was not proactively aggressive, there was little doubt that he would use defensive measures to make me back off whenever he felt scared. Simply crossing my legs was enough to trigger him to bark and lunge at me, and even I find it stressful to be continuously threatened – it triggers my amygdala, too!

Our first strategy was a rather unusual one. Many traditional trainers would have tried to intervene using punishment or a 'startle' technique, such as throwing a can of pebbles at the dog, but it was clear to me that Dave needed to relax rather than feel more threatened. Inspired by the work of 'Tellington Touch' practitioners, who use gentle 'body work' and very specific massage-like touches to calm their animal patients, I sometimes use close-fitting fabric wrapped around the dog's body in order to give them a sense of security. In the past, I have used bandage-like 'body wraps', or have taken a trip to the local charity shop to buy kids' T-shirts. If you get the right size, the dog's front paws go through the armholes and, after a little tailoring with a pair of scissors, the T-shirt can fit over the dog's head and round the chest quite snugly. Nowadays, you can buy actual doggie body wraps designed for this purpose, rather than as a fashion accessory – and their effect can be pretty remarkable.

The way in which body wraps or close-fitting fabric seem to calm anxious or frightened dogs is still up for discussion. It's possible they alter the dog's awareness, so that it focuses on its own body rather than external sights or sounds, or they may simply give the dog a sense of security. Either way, they have proved very useful in many cases, especially when the dog is afraid of fireworks or thunder – and with no side effects, they are sometimes worth a try. Dave certainly looked smart in his body wrap and seemed calmer when wearing it when in stressful situations.

Next, we needed to establish a base-line of security for Dave where he would feel safe enough to eat. Eating is important because the very act of chewing seems to calm dogs (part of the reason why anxious dogs chew up the furniture when their owners leave), and the pleasure of eating tasty treats can reward good behaviour. However, offering a fearful dog a dry dog biscuit is a little like offering me five pounds to go bungee jumping. Quite frankly, I wouldn't entertain the idea for anything less than £50,000.

For this reason, I had a bag full of home-made liver-cake ready and waiting. However, there was no way I could simply put my hand out and expect Dave to take a treat from me. He was far too fearful. Instead, I asked George to take the treats and open the bag. Dave did what any sensible dog would do and turned his nose to sniff the delicious aroma.

That was good enough. Using a clicker, George marked the head turning and sniffing behaviour, then offered Dave a treat from the bag. Like many German Shepherds, Dave was immediately suspicious. He refused the treat and it dropped to the floor. I told George to leave it there and to show Dave the bag full

Liver-cake Recipe

The ultimate training treat – no dog will be able to resist your charms when you carry a pocketful of liver-cake!

1lb (450g) liver (lamb's or pig's)	1 teaspoon of oil
2 eggs	1lb (450g) granary flour
2 cloves of garlic (optional)	A dash of milk

» Liquidise the liver with the eggs, milk, oil and garlic in a blender.
» Add to the flour and mix.
» Put into a microwavable dish and cook in the microwave on full power for 6–10 minutes. The cake should bounce back when pressed lightly, when cooked.
» Cut the cake into slices and freeze. Take out of the freezer when required and defrost before use.

(Note: for dogs with sensitive tummies, you can substitute a tin of tuna for the liver.)

of goodies. Once again, Dave moved his nose a couple of centimetres towards the food. George clicked again. This time, he didn't attempt to feed Dave the treat he had earned but instead dropped it onto the floor next to the other one. Dave sat back slightly. He looked at George, then at the two treats on the floor.

I could almost see his brain whirring. He looked back at George, then deliberately moved his nose again towards the bag. He got his click and the food treat went on the floor. Dave just couldn't resist the number of treats on the floor. He bent his head slowly, very slowly, and sniffed them. George clicked and added a fourth treat to the collection. Dave put his tongue out and tasted one of the bits of liver-cake. Click, and another appeared. He took one into his mouth. Looking for all the world as if he were being poisoned, he ate it very tentatively. I tried not to look at him in case it put him off. Deciding that he might live after all, he ate another.

Now that we had a little window of opportunity, it was important to use it wisely. I gave George a running commentary on Dave's body language and facial expression – which I was checking out of the corner of my eye – as we started, very slowly and deliberately, to click and treat Dave for any hint of muscle relaxation, facial softness, or averted eye contact.

It is well known in human psychology circles that body language, physical movement and even posture can directly affect our emotional state. In a rather neat experiment by researcher Fritz Strack in 1988, subjects were asked to rate how funny they found cartoons while they held a pen in their mouths. Participants consistently rated them as more humorous when they held the pen between their teeth, an action that forced their mouths into a semi-smile, than they did when they held the pen in their lips, which forced a partial frown. This study has since been replicated several times, all with the same fascinating results: facial expression can affect mood, rather than just the other way around. Still sceptical? Sit in a slumped position and hunch your shoulders forward. Let your head droop towards your chest. Sigh deeply once or twice. Look at your feet. How do you feel now? Perhaps it's no accident that we describe the feeling this posture can engender as being 'down'. Now, change your posture and see how it changes your state. Just try to be depressed while standing up tall, clapping your hands in front of you, smiling and keeping your shoulders back. Now look up and to the left. Walk about briskly. OK, you may feel a little silly, but humour me – the chances are you will feel much more cheerful.

Fearful dogs tend to keep their heads still while moving their eyes to follow anything they think may be threatening, so George clicked for Dave's head turns, no matter how small. Scared dogs keep their ears pinned back to their heads, while relaxed and confident ones allow their ears to be in a relaxed but alert position, so George clicked ear movements too. Tension nearly always causes dogs to shut their mouths and hold their breath, or to stress pant – a bit like human hyperventilation. George clicked and treated Dave for a relaxed mouth and, when unsure what else to reward, simply for eating in the presence of someone scary.

In that first session, all I wanted Dave to learn was that good things can happen around a stranger. He had little idea why he was being clicked and treated, but he came as close to enjoying an outing as he ever had before, and despite the fact that I couldn't risk getting up to see them out, I was pleased with our gentle progress. George went home armed with Dave's body wrap and clicker, with instructions to reward calm and quiet behaviour whenever possible.

Over the next two months, George and I worked with Dave several times a week. Gradually, very gradually, the big dog began to relax in my presence and accept me moving about close to him. He would still startle if surprised by a sudden movement, but given enough space he would choose to back away from me rather than attempt to get me to move.

However, we still needed to give him some different options when meeting other people.

George and I started to watch Dave for unconscious reactions when he felt fearful and defensive. We videoed his behaviour and watched it repeatedly for clues. The most obvious of these was a tendency to move backwards one or two tiny paces before coming forwards again in a barrage of lunging and barking. George used the clicker to mark the 'backing up' behaviour, and then reinforced or rewarded it with a piece of food. He needed to be pretty accurate, but the exquisite timing of the click allowed this and we soon began to see results. Indeed, by the end of the session Dave was backing away four paces instead of just two. I sensed we might have a chance of encouraging him to choose that option in a moment of panic and sent the pair home to practise once again.

The following week I arranged a home visit to see how Dave and George were getting on. In this new setting, Dave was once again unnerved by the presence of a stranger, and barked at me from behind the safety of a baby gate across the kitchen door. George ignored this completely, knowing that any attempts at intervention or 'discipline' would only fuel the big dog's anxiety. Instead, once Dave's initial fear had subsided, he was brought in on a long line to allow him freedom but also keep him under control. Instead of rushing at me, teeth bared and frothing with saliva, Dave's new-found option kicked in. He took one look at me and gracefully retreated, by neatly reversing out of the room. Encouraged by his own success, and lured by the sounds of George and I chatting and laughing about this new development, Dave soon reappeared – peering round the edge of the door to see what was going on. This of course elicited a click and treat from George, which brought the big dog another couple of paces into the room.

For the next twenty minutes, Dave shuttled back and forth, in and out of the living room doorway, in forward and reverse gears. Finally, discovering that this was rather tiring, he decided to come right in and say hello. This was the break-through we needed. Allowing him to make his own decision about whether to retreat or approach seemed to give him new-found confidence, which in turn helped to bolster his emotional state when he was around new people and new situations. While Dave was never going to be a dog to wear his heart on his sleeve, at least he had a strategy to employ when the going got tough. Quite literally, the tough got going.

TOP TIPS FOR COMMUNICATING WITH YOUR DOG

- Learning any new language takes a little time and effort. Try to think about how your dog is feeling, rather than simply imposing a human interpretation.

- Basic play gestures, such as the play bow, in which the front end is held in a low stalking posture and the dog's bottom stays in the air, are easy to spot once you know what to look for.

- Dogs clearly experience emotions, but don't be fooled into thinking they are the same as ours. For example, most dogs that appear to be looking guilty are really showing fear.

- Watch out for stress symptoms in your dog, especially in new situations or those that could present a risk – such as around children. Stressed or anxious dogs may react defensively so be proactive in removing your dog from a situation in which he's showing signs of being uncomfortable.

» Watching video footage of your dog allows you to view in slow motion, repeat clips and to watch without sound – all of which will help you to notice subtle aspects of your dog's body language and facial expression.

» Dogs don't understand human words, so if you shout at your dog it probably sounds to them as if you are barking encouragement.

» Your tone of voice is important when talking to your dog – low tones can sound gruff, while high-pitched sounds can be exciting – but your body language is even more crucial. For this reason, try not to bend over your dog or stand 'square on' facing a dog that is lacking confidence.

» Dogs watch our body language and facial expressions avidly. They can easily tell when we are engaged with them or not, and can be encouraged or intimidated by even small changes in our posture and movements.

» Dog wraps and T-shirts can be a helpful tool when treating dogs with fear-based problems – however, in my opinion, that's no excuse to dress a dog up like a human just for amusement. Dogs definitely look and function best in their own 'ready-made' outfits.

All too often, dogs showing aggression are labelled as dominant. Their owners are told they have no control because their dogs lack respect for them, and that they must re-establish their leadership in all manner of ways that domestic dogs are meant to understand.

When I see a dog that is showing aggression, however, I take a different route. I look for the underlying emotional state – and this, in the vast majority of cases, starts out as fear. Fearful dogs would always rather avoid confrontation. They don't want to take risks or escalate the threat they are experiencing. It's dangerous. Avoidance is not possible in many situations – we block dogs' opportunities for flight by having them on the lead, in a kennel, or tied up – and this means they are effectively forced to take defensive action.

Once this has happened and the behaviour has been reinforced, or rewarded, by relief and success, then of course it is going to happen again and again. Unfortunately, for some dogs this new strategy is enough to alter their emotional state from one of fear to one of satisfaction. Now we have a dog that knows how to use aggression and enjoys it. That's a whole different can of worms. The dog that had no way of coping has developed a strategy that works for him. Sadly, it is one that rarely works for us and in the worst-case scenarios the dog can no longer be kept as a family pet.

With this in mind, our mission should surely be to focus on prevention, rather than struggling for a cure. Puppies come into this world with a whole set of genetic potentials, and some of those will be connected to just how well they cope under pressure. While we would like our dogs to live a wonderful, stress-free life, the reality is that this can't always be the case. Every day, I have to get out of bed too early, get ready in no time, go in the office, deal with e-mails, phone calls, traffic and technology, fix the printer, tussle with the mail, work my way through a hundred little annoyances – and that's all on a good day! Dogs also have to deal with life as it happens, warts and all – and how they learn to do this, how they learn to build effective coping strategies, is largely up to us.

3
Team building

Stress immunisation for your puppy

Case history: **Amber, the cotton-wool Cocker Spaniel puppy**

One of the most potent arguments that the old-fashioned 'pack' theorists rely on is that in order to live together, dogs – and, by default, people – must fit into a structured hierarchy. This notion was based on the work of a Norwegian zoologist called Thorleif Schjelderup-Ebbe in 1921. He looked at social systems amongst hens and developed the idea of a 'pecking order' – a hierarchy based on physical dominance, in which one hen would peck another in order to establish rank. The phrase 'pecking order' has become commonplace in everyday parlance in this country and many others to describe social hierarchy. However, it takes more than a single step to extrapolate chicken behaviour to that of the wolf or dog (or even humans), and this is where myth and reality part company.

Watch wild dogs hunting as a pack, and what you see is not a rigid hierarchy at work but a fluid and flexible team operation. In any group of wild dogs there will inevitably be some individuals that are particularly fast, light on their feet or agile. These may be the dogs that chase the prey animal to tire it, effectively corralling it towards other members of the group that have different but no less impressive skills. For example, there might be one or two dogs that are recklessly fearless, and these are the ones who get the job of hanging onto the prey's nose until other heavier or stronger team members do what they are good at and bring the prey down. In such an efficient hunting team, no individual has supremacy over any other; in fact, each individual has an equally important part to play in their survival.

Ah, but what about competition over resources, I hear you ask? What about those classic wildlife documentary scenes where you see two wolves – or even a group of adolescent youngsters – wrestling over a piece of hide or the last bone from the kill? Surely hierarchy has a part to play there? Well, in my view, only humans would watch a group of dogs tussle over a remnant and instantly come to the conclusion that they are competing. What about the possibility that they may be co-operating to rip apart pieces of carcass that are impossible to tackle alone? How about the idea that they might be gathering information about each other? Dogs that rely on a team to hunt need to understand each other's strengths and weaknesses. They need to know whether one individual is faster, stronger, slower or weaker on the right side or the left, and the time to find this out is not at the moment when an extremely angry warthog is bearing down on you, but well in advance during everyday interactions.

Of course, many dog owners find the idea that dogs are really wolves in disguise appealing. It's fun – and rather powerful – to imagine that somehow humans took wolf cubs, raised them in their caves and 'created' domestic dogs. The myth says that we then managed to manipulate how they look and act, breeding them for long coats, short legs and droopy ears, and as long as we maintained 'alpha' status then we remained in control. The myth is wrong, though. As I mentioned in Chapter 1, domestic dogs are not the same as wolves. Despite sharing the vast majority of their genes with their cousins, they are simply not the same creature, as some wonderful 1960s studies demonstrated rather neatly.

In 1959 Dmitri Belyaev, a Russian geneticist, launched a long-term experiment to tame foxes with the initial aim of making them easier to handle for the fur trade. While this may seem horribly unethical to us now, in those days it was essential that captive animals bred for their fur were easy to care for and manage – primarily because injury resulted in the potential loss of the pelt's commercial value. Starting with a population of caged wild foxes, which demonstrated typical fear and aggression towards humans, Belyaev selected cubs from each generation based on one criterion only – those that were tamest around people.

Changes began to appear very rapidly. It took only six generations of breeding for the foxes to start showing friendly behaviours, such as approaching when their keepers arrived rather than running away. Even more amazing was that after only thirty-five generations of breeding for friendly temperament, Belyaev's foxes began to act like domestic dogs. The foxes wagged their tails when they saw their human carers approaching, whined for affection, used appeasement signals and made care-soliciting gestures. However, what was remarkable was not that Belyaev succeeded in breeding friendly foxes that seemed truly to like human contact; it was that with those behavioural changes came unexpected physiological ones too.

The friendly foxes lost their pricked ears and developed floppy ones instead. Their coats changed and acquired black and white patches, like a Collie, and became long and plush. Their tails turned up at the end, like a dog's, rather than hanging down like a normal fox's brush. In addition, the females came into season twice a year, like a bitch, rather than once a year, like a vixen. The foxes also started to bark in a way that was quite unlike anything the keepers had ever heard from a fox before.

Clearly, the genetic shift that caused 'domestication' in these foxes also had many other effects, which resulted in dog-like characteristics. Perhaps, as some eminent ethologists such as Ray Coppinger believe, a very similar set of circumstances occurred among wolves. Some of them might have shown less fear of humans way back in our collective past, and they were the ones that were inevitably 'selected' for breeding by the people who lived close to them, and probably used them as a food source too.

Such 'domesticated' attributes are certainly abundant in the juvenile and social dogs that we keep today. Artificial selection for appearance has done the rest, creating dogs as huge as the Great Dane and as diminutive as the Chihuahua – but this has only occurred relatively recently in dog terms. However, although domestication may have taken the adult wolf out of the dog, it's important to understand that the dog itself is not fully 'tame' unless we help to make it so.

My mother, a primary school head teacher for thirty years, always stood by the adage, 'Show me the boy before he is five, and I'll show you the man'. While genetics have a huge part to play in canine behaviour traits, there is little doubt that the early weeks of a puppy's life are also integral to the way the dog will behave as an adult. If you deny a puppy the chance to meet other dogs, people and the outside world, you can end up with a dog that is effectively institutionalised and fearful of all new experiences.

Puppies who are not exposed to all the sights, sounds and smells that life has to offer before the age of twelve to sixteen weeks may never gain confidence in later life, and may always have problems relating to other dogs or humans. This makes sense. Keep a child locked away in isolation until he or she is eight years old and we would not expect him or her to make a quick and easy social recovery. In fact, we would expect that he or she would be affected for life.

It's obvious that puppies should have lots of positive experiences in those early days and weeks, but they also need to experience 'real life' in a gentle way too. Only by experiencing different emotional states can dogs learn to cope with them, and this means dealing with negative emotions as well as positive ones.

The first time anything negative happens in a puppy's life is at about four weeks of age. Up until then, everything that it needs and wants has been supplied by its mother. Warmth, food in the form of milk, protection, even going to the toilet is prompted by mum – who licks the puppies' bellies to stimulate them to urinate and defecate. Then, gradually, the puppies begin to grow teeth. These are small and sharp and now when they latch on to their mother to feed, they hurt her! This produces an important reaction – mum starts to say no. For the first time in their lives, the puppies are denied something they want. Every time they see her, they clamour to feed, but while she will still allow some feeding, on other occasions she will turn around and walk away from them. Soon, she will also walk off during feeding, leaving puppies to drop off her teats unceremoniously as she goes. As time goes by and the puppies start to develop more muscular co-ordination and the ability to move more quickly and determinedly, mum may have to step up her rejection techniques. She might fix them with a direct stare followed by a deep growl or even a snap or nose-butt if the hungry pups don't back away. In this way, the puppy learns what hard stares mean. Here we should also explode the myth that mother dogs shake puppies by the back of the neck to discipline them. Picking up and shaking only has one purpose in the canine world: it's a killing mechanism (and one that nearly every dog owner is familiar with, as dogs play at 'killing' their toys as part of an enjoyable game). It's clearly not a maternal gesture.

Over time, puppies learn to control their own impulses to rush at their mother and mob her in an attempt to get a feed. In a wild situation, the mother and other adults would bring solid food to the pups via regurgitation, thus successfully redirecting their attention from teats to mouth. For this reason pups still want to lick at our faces and mouths. Domestic dogs rarely regurgitate for their puppies – yet another link with their wild cousins which has been diluted by social evolution. However, it's at this stage that humans start to take over the parental role and supply solid food in a dish to take over from where mum left off.

The whole weaning process is the pups' very first lesson in how to cope with that most difficult of emotions – frustration. Of course, it won't be the last time that puppies experience this. Living with humans exposes them to frustration every single day. In order to get used to it, puppies between eight and eighteen weeks should get out and about to meet and mix with as many other dogs, people, sights, sounds and smells as possible. This builds confidence and reduces anxiety, but it also buffers the puppy for the fact that not everything is going to go their way, not everyone they meet is going to be lovely, and not every dog is going to want to play. Getting this in perspective is basically a numbers game. Venture out for the first time at fourteen weeks old and bump straight into a grumpy adult female Dobermann, and the pup might be forgiven for believing that all other dogs are like this – and learn to avoid them. Meet fifty dogs – males and females, young and old, some lively and playful, some snappy and irritable and some indifferent, and the pup's view of the world will be far more balanced.

A local vet gave me a call to say she wanted to refer a client for behavioural help. The client had been in that morning for her dog's first vaccination, but it had not gone well – indeed, the vet was now sporting a plaster on her wrist where she had been bitten. The bite was quite nasty, and unexpected – primarily because the dog was only fifteen weeks old!

Amber's owner, Tina, called me soon after. She said that she was shocked, not because her beautiful puppy had bitten the vet but because the vet thought she needed behavioural help. In her view, the vet must have really hurt the puppy to make her bite. We arranged to meet.

Amber was indeed a beautiful puppy. The Cocker Spaniel lay cradled in her owner's arms as I was led into the hallway of her new home. I put up a hand to touch her but she turned her face into the crook of her owner's arm and trembled with fear. At fifteen weeks of age, this was not a good response to the careful hand of a stranger. Puppies should be outgoing, curious and friendly, not fearful and withdrawn.

We went into the lounge and Amber's owner placed the puppy carefully on a fleece blanket next to her on the sofa. The puppy glanced at me over her shoulder then slunk down, tucking herself behind her owner's back, clearly hoping that if she couldn't see me, I wouldn't be able to see her.

Amber had come home only a few days before. Her new owner had chosen the breeder carefully and she showed me her puppy's pedigree forms with reverence.

'She has bred Cocker Spaniels for years,' Tina told me proudly. 'I saw Amber's mother and grandmother – they were all stunning. Her grandmother was a champion, you know.'

The puppy had crawled round behind her back and was now heading towards the edge of the sofa. Tina jumped up and scooped it up in her arms, lest the puppy should get too close to the edge. 'The breeder told me how fragile puppies are at this age,' she said. 'She kept them on their own in a warm, padded box, and wouldn't even let other people handle them. They're just too precious.'

Indeed, I thought.

'I think she might need to go to the bathroom,' Tina said suddenly, and carried the puppy towards the kitchen. Following, I expected to see Tina heading towards the back door but instead she turned off down the hallway, and then – to my surprise – took her into the downstairs loo.

There she placed the puppy on a special 'housetraining mat' to do her business. Now, while these 'flat nappies' have become very popular with new puppy owners because they mean that the pup doesn't have to go outside, they effectively condone indoor toileting. This means that owners often need to housetrain their puppy twice: first to the mat, second to the garden. Worse, in my opinion, is that using puppy pads may limit a dog's experiences and mean that it risks being under-exposed to the world at large.

Some years ago, I appeared on a national TV chat show where the topic of discussion was whether people who live in apartments or flats should ever have dogs as pets. Most of the experts on the show condemned the idea of keeping dogs in high-rise accommodation, saying that they need space both indoors and out. I stood out as a lone voice. As someone who had been one of those city-dwelling owners who lived happily and responsibly with a dog in an upstairs flat, I felt I could speak from experience. There are actually some positive behavioural advantages to raising a pup in an urban jungle. Not least of these is the fact that when you don't have a garden or yard, you are forced to take the puppy out and about to meet the world a minimum of eight times a day just for him or her to go to the toilet – potentially seven times more than a puppy living a country existence.

Poor Amber. With so little in the way of life experience and such a sheltered start, she had no coping strategies to fall back on when life threw a minor glitch in her path – in the form of having an injection – and she had over-reacted horribly as a result. The harsh fact is that dogs of all ages need to learn how to cope with being examined, having their teeth cleaned, their nails clipped, their ears inspected, their tails held. They need to put up with minor discomfort in the form of injections, anal gland emptying, temperature-taking and a multitude of other common procedures. All of these trivial little annoyances need to be accepted, not fought over or fussed about, and it is only through extensive amounts of handling, exposure and repetition at an early age that dogs learn to take them in their stride.

At fifteen weeks old, Amber's reactions to other people already represented a behavioural emergency – but getting Amber's doting owner to see that she needed to loosen the apron strings and let her little dog stand on her own four feet was going to be tough. Knowing that the best understanding always comes from experience rather than explanation, I invited Amber and her owner to attend a 'puppy nursery class' that my practice colleagues and I were running during the evenings in a nearby school. I told her it was an ideal opportunity to meet other puppy owners in a gentle social environment. It was also a dramatic eye-opener. On Amber's first session, she sat on her owner's lap and hid her face, not once so much as glancing at the other puppies who were quietly practising training, sniffing each other or enjoying short play sessions with each other.

'Is your puppy ill?' asked a little girl who was there with her family training their Cairn Terrier pup. 'Why can't she come and play with the other puppies?'

Amber's owner looked at me with tears in her eyes. 'None of the other puppies are reacting like this,' she whispered. 'I hadn't realised how painfully shy she is.'

It was a turning point. Very gently, very gradually, Amber's owner tentatively allowed her puppy to explore the house, and then the garden. She practically had to sit on her hands not to dash over and save the puppy from clambering down her six-inch-high rockery. She took her out in the car, let friends touch her and hold her – and even on one occasion left her overnight with a friend (although she did admit to calling almost every hour). Amber's transformation had begun. The following week when she returned to the puppy class, Amber sat on the floor – although admittedly under her owner's chair. She managed a sneaky sniff of another pup's tail as it walked past and even ate a treat given to her by the little girl with the Cairn Terrier.

By week three of the course, Amber could practise 'sits' and 'downs' with the rest of the class. She couldn't yet cope with playing or walking on the lead in the middle of the room, but she wagged her tail and looked more relaxed than I could have hoped for. By week five, she was offering play bows to the

Cairn Terrier and had made a friend in a Bichon Frise puppy who was also a little shy. Everyone could see her progress now.

On the final night of the puppy course, Amber's owner arrived with her pup on a new pink collar and lead. She walked into the room with confidence, sat down and watched as her puppy was happy to be petted by the little girl and a friend who had come to watch the puppy 'graduation' ceremony, in which I say a few words about how each puppy has developed during the six-week course and comment on their achievements. When it was Amber's turn, I hardly needed to remind the class how much more confident Amber had become. They burst into a spontaneous round of applause in their genuine desire to congratulate her owner for all her efforts. They too could see that a crisis had been averted.

Packing up that evening, I was surprised to hear a voice behind me. Amber's owner held out a brown paper bag.

'I just wanted to give you something,' she said. 'You know, for the teacher, from Amber and me.'

I opened the bag. It contained a round and shiny apple. It was undoubtedly the best I've ever eaten.

TOP TIPS FOR 'STRESS IMMUNISING' AND SOCIALISING PUPPIES

» Start young. Even if your puppy has not yet completed all his vaccinations, you can carry him out and about to meet the world. The first critical window of opportunity for puppies to learn to cope with everyday life is before twelve weeks; after this, every day becomes potentially more difficult.

» Try not to wrap your puppy in cotton wool. He or she needs to learn how to cope with life. The balance between protection and exposure is an important one.

» Dogs need to be exposed repeatedly to all the sights, sounds, touches, smells and even tastes of their environment. Treat your puppy as if he is going to be a guide dog by taking him out and about as much as possible.

» Be brave enough to leave your puppy home alone for short periods.

» Find a good puppy class and enrol your dog as soon as they allow. The class should be specifically for pups of eighteen weeks and under, and should offer a combination of carefully controlled socialisation with other puppies and kind, gentle training.

» Even very young puppies can sometimes show problem behaviours. Don't be fooled into thinking that he or she will grow out of it. Seek help early if you need it.

» Puppies and children are a wonderful, if sometimes wild, combination. Make sure that both have 'calm down' periods and that your puppy has somewhere he can go and rest undisturbed. Never leave your puppy unsupervised with children.

» Pups of eight weeks old can learn basic manners and training, such as 'sit', 'down', 'come when called' and multitudes of simple tricks such as 'spin', 'rollover' and 'give a paw'. Learning is easy and fun when you are young.

» Puppies often go through a 'fear period', characterised by being confident one day and then being scared of something commonplace the next. Don't reinforce the fear by giving attention; instead, pretend nothing has happened, wait until your pup recovers, then reward him for being brave the next time he encounters the same thing.

» Introduce your puppy to good-natured adult dogs as soon as you can. A 'telling off' from an older dog may be perfectly appropriate if your pup is too bumptious. This is distinguishable from aggression as it is all noise and bluster with no risk of damage. Don't panic if it happens, as it's likely to do you a favour.

» Two puppies from the same litter need to be walked, trained and socialised separately if they are to develop as individuals in their own right, and not many owners have the time or dedication needed for this. Pups that are over-dependent on each other or an older dog in the same household run the risk of lacking real-world experience and may have problems later.

Amber's story is not uncommon. Her initial inability to cope with life outside a very small, protected world was caused by a complete lack of experience in those precious, formative weeks. This gives clues as to how pups should be raised once they are in a new home, but also what should happen while they are still with their breeder, their litter mates and their mum. Pups need a balance of protection and stimulation, of security and gentle exposure. Such a tightrope cannot be carefully negotiated if the pup is over-protected, shielded from everyone and everything.

For those that are born in a barn, surrounded by the barking of other dogs, with only the warmth of a heat lamp and the feel of shredded newspaper for stimulation, the outcome is strangely similar. Sadly, puppy farms are still common in the UK, with enough prospective owners willing to take on dogs that have been bred and raised in social deprivation, despite all the problems that we know this can cause.

A pup's early experience can make or break its chances of becoming a family member. Of course, it's hard to turn down the kids' pleas and say no to a pup, even if you know it hasn't been reared in the right way. Sometimes it is precisely because you know that the conditions are poor that you want to 'rescue' the pup and offer it a home. Dogs have a way of getting into our hearts like no other animal. They compel us to take them, to care for them and to spend inordinate amounts of time and money on them. All those songs are right: love makes us crazy – and it can make our dogs a little bit daft too.

Part 2

Love

4

Emotions and addictions

Dogs who love too much

Case history: **Sam, the besotted Springer Spaniel**

Going out to work if you keep dogs has always been controversial. Clearly, it's not feasible to keep a dog if you are out for eight or ten hours a day. Dogs are social animals, and even if they are given food, water and shelter, their emotional welfare demands that they have exercise, company and stimulation during the day. However, it's an ironic fact that in order to keep pets, most of us have to go out to work to be able to buy their food, pay their vet bills and buy them all the many beds, treats and little extras that make them (and us) happy.

Like most things in life, achieving harmony is all about balance. It should be perfectly reasonable to leave your dog at home for a realistic period of time. Of course, what is a realistic period of time will depend on both dog and owner. For example, a puppy of only twelve weeks cannot be expected to go more than two hours without needing to go out to the toilet and have some meaningful interaction. Dogs over the age of a year who know the routine and have had exercise and stimulation may be OK left for four or even five hours, especially if their owner is savvy and leaves them with plenty of things to do in their absence. In fact, I have treated some dogs for behavioural problems that have been caused by owner over-obsession. Some of them would probably have paid me themselves if I could have persuaded their owners to go out and leave them in peace for a couple of hours! It's when the balance goes wrong that I sometimes find my phone ringing, as I did with Sam.

Sam was a healthy, typically energetic two-year-old Springer Spaniel. He loved to hunt for imaginary quarry in the fields near his owner's home and chase his ball in the local park. He also loved his owner. In fact, he loved her so much that he simply couldn't bear to be parted from her.

Sam had been given to a rescue centre when he was about thirteen months old. His previous owners had not been able to manage his energy levels with three small children to look after, and so, through no fault of his own, he found himself looking for a new home.

Karen had wanted a dog for some time. She had previously owned dogs when she lived in her parents' home and since she had gone part-time at work, she was keen to add another four legs to the family. Sam came home and immediately bonded with her, following her from room to room and even wanting to accompany her to the loo. All was well until Karen's long-term boyfriend decided to leave. Upset and distraught, Karen turned to Sam as her only real companion, comfort and confidante. In her words, they became 'inseparable': he was her rock, her reason to get up in the mornings. Then, due to increased financial pressures, Karen had to go back to work full-time – and Sam's world fell apart.

Karen organised a dog walker to come and take Sam out at lunchtime once she was back at work, and he clearly enjoyed running in the woods for an hour with two other dogs and generally having a good time. However, during the afternoon Sam would react badly to being left, and this is where the problem began.

In the beginning, it was apparent that Sam was relieving his anxiety with a little recreational chewing. First, cushions began to take the brunt. Karen would come home and find them shredded and flung to the four corners of the living room. She cleared up the mess, ticked Sam off, and thought little more about it. However, over the next few weeks, things deteriorated, and it wasn't long before she dreaded coming home and putting the key in the door, waiting with bated breath to see what destruction lay before her. With the cushions destroyed, Sam had moved onto the other soft furnishings. The sofa, a patch of carpet and even the curtains were all scratched, torn down and chewed. Most of Sam's efforts seemed to be concentrated around the doors and windows – indeed, the slobber marks on the glass told a story in themselves – and it wasn't long before the skirting boards around the hall door began to look as though they had been put through a shredder.

In desperation, Karen decided to leave Sam confined to the kitchen. This seemed to work for the first few days, but then he turned his dental attention to the cupboard doors, the lino and the back door. The level of his destructive capabilities knew no bounds, and within three months he had managed to do an impressive demolition job on the kitchen, amounting to thousands of pounds' worth of damage.

Of course, like any reasonable human being, Karen was both upset and angry about Sam's actions. She tried telling him off, dragging him to the damage and pointing at it in order to try and make the association between his behaviour and her annoyance, but if anything, this made matters worse. Running out of both patience and money for repair bills, Karen tried leaving Sam in the garage, where his damage options might be more limited, but was horrified when she came home to find that he had spent the afternoon attempting to chew through the wire which held the up-and-over door in place. The wire was intact but his mouth needed veterinary treatment due to the cuts on his lips and the injury to his gums. In a final act of desperation, Karen followed a friend's advice and tried leaving Sam the free range of the house, in case he was fearful of being shut up in one room. That night, the sight she came home to looked a little like the aftermath of a scene from *Lethal Weapon*: Sam had rampaged through the house, shredding her duvet, destroying any object that he could get hold of, and finally – in a sequence of which any stunt man would have been proud – had quite literally jumped through the glass of her upstairs bedroom window and landed in the front garden.

Remarkably, Sam survived this amazing feat with only a few minor cuts to his paws, thought to be the result of pacing around on the broken glass after he landed. Thinking that the house was being burgled and that the dog had been thrown from the window, Karen's neighbours called the police, who were mystified to find that there had been no forced entry – only a dramatic exit.

Dogs clearly show an emotional range that equates to ours. While even ten years ago it was frowned upon to talk about animals' emotions, in these days of neuro-scientific advancement we can talk about the fact that dogs experience, and express, a range of emotions that we can empathise with. However, there are species differences, and this is where confusion often arises. Dogs clearly show sadness, anger, joy and attachment. However, they are also more simplistic and 'honest' in the way they will attempt to redress the balance of their own emotional states. Just like a depressed human, a dog may become 'addicted' to eating or chewing in order to elevate its mood, and clearly this can get them into trouble when the chew toy is the skirting board or the interior of the car.

Poor Sam was not trying to get revenge for Karen's absence. It may sometimes be tempting to think that a dog is getting even for being left on its own, but this is never the case. Most people have heard of the expression 'separation anxiety', but there are many reasons why dogs can become destructive when they are left alone and some are more obvious than others. Bored dogs may simply wait their chance to have a really good chew when their owners leave the house. I call this 'separation fun'! It's simply an under-exercised or under-stimulated dog occupying himself with the equivalent of a doggie PlayStation. However, others are genuinely suffering from a form of separation disorder – and of these, separation distress can be one of the most dramatic.

One of the very first things I do when faced with a dog that cannot cope alone is to find out what's really going on when the owner is not there. Leaving a camcorder running to find out what your dog gets up to in your absence is always fascinating and often hilarious. I once caught my own cat on camera, sitting on my kitchen work surface, looking as smug as any cat can while he happily patted frozen chicken joints (which I had left out to thaw) down into the open jaws of my Golden Retriever waiting below.

Sam's behaviour on tape was exactly what I would have expected for a dog with genuine separation distress. Even before Karen left the house, he was in a state of high anxiety: panting, pacing, starting to salivate, and frantically following his owner from room to room as she put on her work shoes, picked up her keys and checked that the back door was locked. Dogs are good at reading signals. After all, they have many idle hours a day in which to watch them and analyse the habitual patterns of human behaviour. Sam knew that all these actions pre-empted his most dreaded experience – being left alone – and his stress levels went up a notch with each added signal that confirmed his worst fears were about to be realised.

Once Karen had left the house, Sam tried a few barks and howls. Standing by the front door, he threw his head back to howl, then stopped to listen to find out if his owner would respond to his cries for help and return home. When this didn't work, he ran frantically from room to room, checking that she really wasn't there. He jumped onto the sofa and looked out of the windows, leapt onto the front windowsill and – as if in a frenzied trance – scrabbled madly with his front paws at the sill and window surrounds, using his teeth on the handle in a frantic attempt to escape.

At this point, Karen came home. For the purposes of the video, I had told her to leave the house for just a few minutes. After all, we wanted to make sure both her dog and her furniture were still intact on her return. As soon as Sam heard Karen's footsteps on the path, his behaviour changed from one of intense anxiety to horrible internal conflict. Here was his beloved owner, with whom he was desperate to make contact, but also present in his mind was the dread of uncertainty. Just how would she react when she came in? Would there be joy at the reunion or shouts of anger?

Anyone who has ever worked for an unpredictable boss will know a little of what Sam was suffering. Recall that horrible memory of a miserable Sunday night, the churning stomach, the feeling of dread – all because you have to face them again on the Monday morning. Just how will they react? Will they have had a good weekend and be all smiles and chats over coffee, or will they be stony-faced and grim as they call you into the office for a dressing down? Such behaviour has all the hallmarks of random emotional punishment – and if humans don't like it, I think it's fair to assume that neither do dogs.

Poor Sam squirmed on the floor as Karen came in. His ears were glued flat to his skull, his tail tucked under, his head hung low. He even avoided eye contact, despite being desperate to reach his owner.

Such behaviours are hard for loving owners to watch. 'Look,' said Karen. 'That's what I mean. He knows he's done wrong.'

One of the wonderful aspects of dog behaviour is that they are always honest when it comes to their body language and facial expressions. Dogs don't fake smiles, nod their heads when they mean no, or tell you that they love you while really having an affair with your best friend. Ironically, this honesty

often gets them into trouble. Owners might think their dog feels guilty because his body language matches our idea of what guilt looks like in a human, but lowering the head, rolling the eyes, pinning the ears back and tucking the tail under are all classic signs of fear, not remorse. They indicate that the dog is anxious about how the human will react rather than worrying about what he has done. No one wants to think that their dog is afraid of them, the person they love most in the world, but to the dog that is happy to see his owner, it comes as a huge surprise when that person is angry and aggressive. They don't make the connection with their own previous actions – whether that be chewing, scratching, going to the toilet or barking – sometimes even when they are caught in the act.

Sam wasn't getting revenge or being naughty when he was left. His addiction to Karen's presence meant that when she left he started to experience emotions that can best be described as going 'cold turkey'. In humans we would recognise this easily – but in dogs? Certainly. Sam's panic at being left alone was only partially relieved by chewing. His drive was still to attempt to escape and be reunited with his only security – his owner, Karen.

On the whole, separation disorders respond well to behavioural treatment. However, practicalities can get in the way – the main one being that while treatment is in progress, the owner still needs to go to work, triggering the original problem over and over. Karen arranged to take a week off work. With two weekends and a public holiday thrown in, that gave us twelve days to make a difference. It wasn't much time, but we had to bank on the fact that Sam was a fast learner.

In the first five days, the main focus of our efforts was based on reducing Sam's dependence on Karen

as his sole attachment figure. We did this by cooling off their relationship a little: getting a friend to come in and feed him and play with him, restricting his access to Karen while she was in the house by using a baby gate, and encouraging him to nap in his own bed in the kitchen rather than on her feet. Like many dogs with an owner-contact addiction, Sam was hyper-aware of Karen's whereabouts whenever she was in the house. He would follow her to the bathroom, lie touching her when she watched TV, and would even sigh and move himself closer if she deemed to move her feet an inch away from him. This impression of 'me and my shadow' had to stop, but it was just as hard for Karen as it was for Sam. She clearly struggled with cooling off her relationship with her dog. In fact, I rather wondered who really had the separation problem.

After a week, Sam had begun to accept Karen coming and going through the baby gate without getting into a panic. Now the time was right to re-direct his addiction onto something else and instigate some coping strategies for when he was on his own.

Just as humans can become addicted to chocolate, drugs, shopping or playing computer games, so dogs can become addicted to certain behaviours. This is well documented in various research papers, although it may not be described as 'addiction' because the authors don't want to anthropomorphise animals. However, in my practice we see dogs who are addicted to barking, chasing, spinning in circles, biting their own tails, chewing their own limbs and even licking themselves to the point where they create open wounds. Unfortunately for such dogs, these behaviours can start to cause neurochemical effects in the brain that are pleasurable and addictive, not unlike the endorphin 'high' we might experience after a run.

For this reason, it is well nigh impossible to simply stop the 'addict's' contact with the focus of their addiction. Instead, we need to form a new addiction to something else – preferably something appropriate. In Sam's case, it was clear that part of his preferred coping strategy was chewing, and as this helped to bring him relief from anxiety it was number one on our list.

Over the course of a couple of days, Sam was shown a hollow rubber 'Kong' toy, which was filled with the most delicious treats imaginable, packed in such a way that getting them out would present a bit of a challenge. There was cheese in the very bottom, Marmite around the inside, and dog biscuits filling up the top. Frankly, this culinary delight was every dog's dream offering. It smelled heavenly (if you were a dog). Although Sam was as keen as mustard to get the toy, he was simply shown it then it was put away. As you can imagine, by the third day, the effect of a little built-up frustration was beginning to have an impact and Sam was desperate to get his paws on it. At this stage, Karen put up a sign in her kitchen. On a large piece of paper was written 'Time out'. This was stuck to the front of the fridge door, and – immediately after – Sam was given the delicious toy, then ignored. He immediately got stuck in, and seemed fine about the fact that Karen was ignoring him even though she was in the same room. Getting Sam 'hooked' on the toy in Karen's presence was an important step, as many dogs are so dependent on their owners that they will refuse to eat when they are parted from them.

After a few minutes we took our chance, and I was pleased to see that Sam barely looked up when Karen left the kitchen, shut the baby gate behind her and went upstairs for a few minutes. Not wanting to push our luck, she came back down, distracted the happy

Sam from the Kong by giving him a treat, and put it – and the sign – away.

This little routine was to become a habit with Sam and Karen over the next few days. Sign up; Kong filled with delicious food down; Karen out of sight and physical proximity. Karen ventured as far as the attic, the bathroom, the spare bedroom and even out the front door, all without Sam so much as lifting his head.

It was good progress, but still not enough for us to be confident that Sam's addiction to Karen's physical contact was 'in recovery'. In 'belt and braces' style, I needed to be sure that Sam had made the connection between the signal and his new coping strategy, that he knew the sign going up on the fridge door meant he simply had no chance of getting Karen's attention, and that he might as well resign himself to lying quietly and chewing his toy.

I left them to do their homework.

Four days later, I called round to see how things were going.

'It's pretty good,' Karen said, as I stroked Sam's silky head. He was leaning against me, brown eyes gently blinking in the heart-melting way that only a Spaniel can. 'In fact, I think he's learning how to read.'

Karen went into the kitchen and opened a drawer. She took out the sign that read 'Time out' and turned to stick it to the fridge. Sam, attentive as always, had followed us in. He took one look at the sign, gave a little snort, and trotted over to his bed. There he sat, waiting patiently, lifting one front paw and then the other in a gentle show of anticipation. It was a million miles away from the emotional state that he had been in before.

'Look at him,' she said. 'I'm beginning to feel a little jealous of that toy! Shall we go out?'

We packed Sam's toy full of tempting titbits and left him. Over lunch at a local pub, we discussed how the new routine was going to be handed on to the dog walker, and chatted about exercise and feeding regimes that would suit this energetic gundog on a long-term basis.

Tentatively, we walked back to the house. Creeping round the side, we peeked in the kitchen window to spy on Sam. Still in his basket, and with his jowls resting on the toy, Sam slept. Karen looked on. I put my hand on her shoulder, and a slow tear slipped down her face. I wasn't sure which of the three of us was most relieved.

TOP TIPS ON PREVENTING SEPARATION PROBLEMS

» Use a baby gate so that your dog can see you but not always get to you when you are in the house together.

» Practise 'little and often' separations right from the start of your relationship with your new puppy or dog.

» Exercise your dog before leaving him, and make sure he has been to the loo.

• Be casual about leaving. Avoid an *EastEnders*-style drama when you say goodbye.

» Be social but calm on your return home. Never, ever punish your dog if there is damage or he has been to the loo in the house. It will make it worse.

» Leave lots of things that your dog likes to chew.

» Your puppy will be distressed when separated from you at night if he has been with you all day long. Make sure he spends some time away from you when he is tired and sleepy during the day too.

» Does your dog want to follow you like a little shadow when you are at home? He's forming an addiction, which could be a problem later. At the very least, shut the bathroom door (with him on the outside) when you go to the loo.

» Be realistic about how long you can reasonably leave your dog. Puppies under the age of twelve weeks should not be left longer than two hours at a time. Adult dogs that are used to a routine can be left for longer, but bear in mind that dogs are a social species that need company, as well as exercise and the chance to relieve themselves, on a regular basis.

» Indoor crates or cages are wonderful secure dens for dogs, but don't abuse them. A dog should not be left in one for more than a couple of hours, except at night. Introduce the crate carefully – and make sure your dog views it as a comfy bed area rather than a prison.

» Finally, if you love your dog as much as I do it's natural to find it hard to leave him, but for your dog's sake, you must. Establishing that short periods apart are normal will give balance to your lives and will prevent trauma when separation is unavoidable.

Of all the useful habits new dog owners should adopt, leaving their pup or dog alone for short periods is perhaps the most important. Dogs need to learn to cope with gradual separation from their owners, and this doesn't always come naturally. If you have the opportunity, leave a video camera running on the area of your home where you think your dog is most likely to spend time when you are out; it can be a source of information – as well as potential amusement. Seeing your dog on camera when you are not there to supervise tells you much. In my years, I have seen dogs that are so distraught their owner has left that they lick and chew at their own limbs, finding solace in the natural opiates that are released in response to pain. I have also seen dogs that champ at the bit, waiting for their owners to leave, so that they can finally have some fun and enjoy themselves, albeit by opening the bin, the fridge or the laundry basket!

Despite our dogs' need to be a part of a social group, and to enjoy the security that this brings, they are hugely flexible in their behavioural repertoire. If you form sensible habits early on, such as leaving them for brief periods, it helps dogs to establish a routine, something they seem to find particularly reassuring. Dogs are dogs, not small people in furry clothing. They need to be able to fit in with our lifestyles, schedules and environments. Giving them clear, consistent signals and realistic boundaries is not an attempt to keep them 'in their place', but rather a way of helping them feel 'in pace' within our families.

JVK

5
Relationships

How your dog sees the world

Case history: **Hector, the little dog with big ideas**

Sharon and George were devastated when their old German Shepherd-cross died at the ripe old age of fifteen. Life wasn't the same. They missed the company, exercise and social opportunities that having a dog brought into their lives. Feeling sad, and on a whim, they took a drive out to a 'puppy warehouse' on the outskirts of their town. What they saw there broke their hearts: puppies of nearly every breed crammed together into stable units, like battery hens, all clamouring and yapping for attention, and leaping up at the half-heck doors fruitlessly trying to get a glimpse of the outside world.

Suppressing their horror, Sharon and George walked down the line of stables and were shown several different breeds of pup for sale by the owner of the place. It looked as though the pups were of mixed ages, and although the breeds were kept in groups together, no mums were visible. In the last stable unit, huddled under a single heat lamp, were some tiny Yorkshire Terrier puppies. No bigger than guinea pigs, the dogs whimpered and shivered in the stark surroundings. Sharon and George lingered at the door.

Sharon counted them. 'Nine,' she said.

'Ten,' replied the proprietor curtly. 'There are ten in there – or at least there should be.' He pushed open the stable door and went in, moving the puppies aside one by one so that he could count them. There, right in the middle, was number ten – the smallest and most pathetic-looking puppy the couple had ever seen. The owner scooped him up.

'Four hundred and fifty pounds,' he said, pushing the puppy towards Sharon. 'Should be six hundred, but it's a bit smaller than the rest.'

Sharon cradled the puppy in her arms, and tried to soothe his trembling frame. He felt chilled and somehow brittle, like a frosted leaf. There was no way she was going to leave him there.

Knowing that they weren't really doing the right thing in buying from a puppy farm, but driven by a heart's desire to rescue at least one of these poor puppies from their impoverished surroundings, George paid up, and Sharon took the tiny puppy back to the car, snuggled under her coat. As she got in, she could hear George's voice raised in anger.

'That man!' he complained as he got into the car with Sharon. 'He says no receipt, and no refund. Says the puppy is registered with the Kennel Club and he'll send on the papers, but then he didn't even want to take our address. It's obvious all he wanted was our money.'

'Poor little mite,' murmured Sharon, as she cuddled the puppy in close. 'I wish we could take them all.'

And so it was that Hector came to live with Sharon and George. In the next few months, like so many other puppy farm puppies, he was a frequent visitor to the local vet, suffering with an umbilical hernia, digestive complaints, ear mites, worms and a nasty skin rash. Research has shown that puppies raised in a socially deprived environment have less resistance to disease, as well as suffering behaviourally. Despite the ever-increasing costs, the worry and the difficulties with allowing Hector to lead a 'normal' puppy life, Sharon doted on him. Indeed, it was this very close bond that made it all the more difficult to admit that they also had an aggression problem to deal with.

Despite his physical weakness, right from the start Hector was a demanding character. He barked to be picked up, didn't like going out for walks and would create so much fuss when they tried to leave him alone that they decided it would be best if one of them stayed at home with him 24/7.

Hector also hated anyone near his bowl at feeding time, and would snarl and show his teeth if either Sharon or George approached. This made them laugh at first; seeing such a diminutive dog transformed into a savage beast was strangely contradictory. However, it wasn't long before Hector started to fly at their feet if they so much as stepped into the kitchen while he was eating. What had seemed funny to start with now seemed like a major problem – after all, even tiny dogs have teeth.

By the age of fifteen months, Hector had turned into a canine tyrant who ruled his owners' lives. He was effectively making this well-intentioned couple prisoners in their own home, and dictating when and where they could walk and sit. He would bark continuously at them if they watched TV instead of giving him attention – and on one occasion even cocked his leg on the DVD player when they were engrossed in a film. His demands to be picked up and cuddled were beginning to have an impact on their relationship – and the fact that he liked to sleep on the bed between them wasn't helping. The final straw came when George got out of bed one night to go to the bathroom, and came back to find that his way into the bedroom was blocked by a seriously grumpy Yorkshire Terrier, who obviously meant business. Admitting defeat, George spent the rest of the night on the sofa, but in the morning he called me.

Despite their similarities to us, dogs are also vastly different. Most obviously, they don't speak English or

any other verbally based language. Although domestic dogs use a range of vocal communications, sound is relatively insignificant in comparison to their languages of sight and scent. Frankly, living with a chatty human is probably like living with the radio on for most dogs. Perhaps it's little wonder that so many appear to ignore almost everything their owners say to them. Instead, dogs make up for this by noticing almost imperceptible visual changes in their environment – and us.

Most dogs can tell at a glance whether we are going to work or preparing to take them for a walk just by the shoes we are putting on. My dogs can tell when I am expecting visitors (perhaps the frenetic vacuuming does give them a clue). Even more amazing are the dogs that can predict when their owner is about to have an epileptic seizure anything up to an hour beforehand. Dogs are the ultimate surveillance experts. They are effectively captive every day of their lives, and this gives them a 24-hour opportunity to watch and learn. Most dogs work out their humans' routine within 48 hours of being in a new home. Within seven days, they have figured out how to break that system quite simply in order to make it work to their advantage.

Once again, this is not dominance. The dogs I see that recognise opportunities to get what they want are not vying for rank or trying to be leaders – they are just bright dogs. It is for this reason that tiny dogs, and runts of a litter in particular, are the most likely to learn how to get what they want and when they want it. They, more than most, have had to learn how to use brain over brawn in order to survive and so are top of the list when it comes to turning a situation to their own advantage. Of course, Hector fitted this bill perfectly.

Many so-called 'symptoms of dominance' in dogs are based on the misapprehension that they behave in certain ways in order to get further up the hierarchy, but in fact many of these behaviours, such as resource guarding, have far less complex causes. All puppies – with the exception of singletons – have to compete with their littermates for food, warmth and human attention. Good breeders recognise this and do everything possible to reduce this competition and maintain harmony within the litter. Wisely, they give the rapidly growing puppies enough space so that they are not overcrowded. They allow the pups freedom of movement and give them things to do and objects to play with, so that they are not simply pestering their mother all the time. They ensure that all the pups have plenty of one-on-one quality time with humans, so that human attention doesn't become a resource to be squabbled over. Most importantly, as the pups are weaned, they provide a high-quality solid-food diet in lots of small dishes rather than just one large one. This is important because give only one plate of food to twelve hungry pups and the biggest, greediest or most aggressive will be the one to eat most, while the rest are forced to compete directly for what is left. Sadly, such behaviour learned as a puppy can continue into adolescence and beyond, and Hector's behaviour in the kitchen and around his food bowl certainly seemed to be showing classic characteristics of a dog that had learned how to keep his food to himself.

Of course, such behaviour is not only effective in keeping food – or a play item – in the possession of the dog that is defending it; the feelings of relief and satisfaction that it can bring also start to have a self-rewarding impact and increase the chance that the dog will repeat the behaviour.

One of the greatest myths that surrounds 'rank order' in dogs is that only the top-ranking dog will possess and defend food and other resources. Some proponents even claim that the 'top dogs' will eat first while the others wait for leftovers. While this may sound appealing as a theory, in practice it is simply not true. Make an important resource such as food scarce and no matter the size, breed, type, age, gender or history of the dog, he or she will defend it. In the world of dogs, possession is definitely 100 per cent of the law. Give a hungry puppy a piece of food and he will guard it from other puppies, adult dogs, even humans if he feels that his survival depends on it.

All too often, I read advice in dog training books on how to prevent a puppy learning to guard its food. Owners are told to take the food bowl away from the puppy while he is eating, in order to get him used to it. This may be well intentioned, but it is the antithesis of prevention – and in fact can often cause the very behaviour problem the owner was attempting to avoid. Imagine you are hungry and looking forward to a delicious meal. You sit down with your plate and have just got stuck in when someone enters the room, approaches you and takes your food away. The person keeps it a while, then praises you and gives it back. The first time this happens, it's likely you are simply surprised. On the next night, you are tucking in when the same thing happens again. This time it's less surprising and more annoying. On the third occasion – humans are fast learners, after all – you hesitate over your food as the person comes into the room while you are eating. This mini-freeze is actually a non-verbal warning, but the other person does not heed it and heads towards your plate with the intention of removing it again. Your response at this stage is likely to involve blocking the person from taking the food – either by questioning them, or by physical means. Whichever you choose, it's likely an ugly scene will follow if the food-stealing individual says something like, 'No person's going to guard food from me', and insists on taking the plate away.

Of course, when this scenario involves a dog and a human we stop looking at it in terms of common sense and start applying all kinds of 'rules' we have read, which we believe should make dogs willing to give up resources instead of defending them.

Want to turn this situation on its head? Well, how would you feel if, while you were eating, someone came into the room and brought you a little extra treat, which they left next to you. If they did this repeatedly, the chances are you would start to look forward to their presence rather than dreading it. Rather than fearing that you were going to lose your food, you would start to anticipate that their arrival meant extras. This strategy works just as well with dogs. Pop a bit more food into the dog's dish, or right next to it, each time you pass by while they are eating their food, and they learn to welcome your arrival and proximity to their food bowl. No stress, no confrontation, no hassle.

To say that Hector was not exactly thrilled by the presence of another human in his owners' flat on the day I visited is an understatement. Although I sat quietly and used non-threatening body language towards him, he barked at me continuously for about forty minutes – quite a feat, as barking is pretty tiring for any dog. Once we were able to get a word in edgeways, George told me that things had gone from bad to worse in the last week while they waited for my visit, and that he had been prevented from getting into bed on two more occasions. Now, as any professional behaviourist will tell you, the devil is always in the detail of such discussions. To be frank, issues with dogs and bedrooms often reveal far more about the owners' private lives than I care to know, but I asked exactly what had happened.

'There was the most almighty hoo-ha in the middle of the night,' Sharon explained. 'I could tell George and Hector were having words and then I heard George stomp off to the bathroom. Next thing I know, it's morning and neither of them are there.'

'No, I wouldn't be,' said George, almost spitting, 'on account of the fact that I had to sleep in the bath all night. Bloody dog wouldn't let me out of the bathroom.' He looked at the dog through narrowed eyes. From the security of Sharon's knee, Hector looked back. He might be small but he was determined, and he had a full set of teeth that would do no less damage than any other dog on a mission.

I turned to Sharon and I couldn't be sure but I thought I caught a hint of a smile. It wouldn't be the first time I have been called out to treat a 'contraceptive dog' – one that won't let the couple sleep together. I usually find this arrangement suits one party more than the other, and I had a nasty feeling that this couple might need Relate more than they needed dog training.

After an hour, Hector was still in no mood to be friendly to a stranger. I tried to endear myself by offering tasty treats, but was told that he would be unlikely to eat any kind of food – even chicken or ham – unless it was served warm and in his favourite dish. He sat on Sharon's lap, shaking with anxiety and adrenaline, while she petted and soothed him and told him how good he was. His behaviour appeared rather less than good to me. I suggested that we took him out for a walk, but was told that he hated walks and probably wouldn't cope with having his lead put on while I was there. George tutted and stared out the window.

Finally, after taking as much useful background history as I could, I asked the couple the million-dollar question: 'What do you want to achieve?' For the first time, they looked at each other and then at me.

'It's obvious isn't it?' Sharon exclaimed. I stayed quiet and raised my eyebrows. 'We just want him to stop peeing in the house!'

Now, it may be an obvious fact, but what constitutes a problem behaviour for one owner may be perfectly acceptable to another. Dedicated dog lovers have been known to exclaim in horror when they see that I allow my dogs on the sofa with me, while I squirm at the knowledge that my friend allows her pets to lick the dinner plates before they go in the dishwasher. Of course, we all differ in what we find acceptable or unacceptable, but just occasionally an owner's expectations may seriously jeopardise human or canine welfare. In these cases, it can be a delicate counselling issue to present the facts as they really are.

From my observations, Hector was living almost as miserable a life as George. While Sharon seemed to enjoy Hector's total dependence and incessant demands, there could be little doubt that the tiny dog himself was full of anxiety and insecurity – and that only by gaining constant reassurance from his owner could he maintain the status quo. Hector was afraid of life, especially anything new, and the only way he could feel safe was to attempt to keep anything novel as far away as possible. Of course, this worked. Hector's barking and snarling originally arose out of fear, as a survival strategy, but quickly became a successful way of making people back off. He had successfully stopped his owners trying to take him outside and was even preventing George from getting too close to his safety blanket, the doting Sharon.

Unfortunately, Hector's background had made such fear and anxiety far more likely than most. His socially and environmentally deprived puppyhood, combined with his many illnesses, had meant that Hector's early experiences of the big wide world had been severely limited. Combined with an exaggerated attachment to Sharon, the rewards that had come from his behaviours in the form of laughter and petting had been enough to set the seal on his present behaviour.

I asked George for his view. He glanced rather nervously at his wife and said, 'That's important, but I'd like to be able to walk around the house without fear of being bitten, too.'

Sharon opened her mouth to launch into a defence, but I got in first. As gently as I could, I explained that I thought Hector was missing out on life. He might get plenty of love and attention in the house but in order to live a full life, he needed to be able to share some loyalties with George, and to enjoy some fresh air. What would happen if Sharon had to go away, perhaps for a stay in hospital, or if she needed to leave the house in an emergency when George couldn't get home?

Sharon was not convinced. Despite my best efforts, I left that day having given her advice on house training to prevent a recurrence of the indoor urination that was top of her list. I still couldn't go near Hector, and as George showed me out, I was sure I heard Sharon reassuring her little dog that the 'nasty lady' was going away.

Despite several attempts to contact George and Sharon over the next few weeks, I heard nothing about their progress – or lack of it. In fact it wasn't until I visited the referring vet practice that I heard news of the couple.

'So much better,' said one of the vet nurses. 'You worked wonders!'

This was news to me.

'Oh yes,' she said, seeing my surprise. 'We saw George yesterday. He moved out, you know. He and Sharon split up. He's so much happier in his new home!'

TOP TIPS ON CREATING HEALTHY BOUNDARIES

» Don't buy a puppy from a puppy farm, or a pet shop – which often tend to be supplied by puppy mills. Some outlets for puppy farms look remarkably attractive – the best bet is to always see the pups with their mother. Don't forget your puppy will be influenced by its genes.

» Dogs can be spoiled. If you treat your dog like a baby or a child, you are preventing him from expressing natural canine behaviour as well as creating potential behaviour problems for the future.

» Just because a dog is small doesn't mean he is a cat. Small dogs often have big personalities and they need just as much early socialisation and training as big ones.

» No matter what shape or size, a dog is a dog. They have legs and should be allowed to use them. Being carried in a designer handbag or picked up as soon as another dog appears is completely contrary to any dog's natural behaviour.

» Play with your dog as much as you can, using toys to encourage him to seek, find, retrieve and drop. Those that play together build strong bonds together.

» Be parental by creating boundaries and sticking to them. It's far more important to be consistent, firm and fair, than it is to be dominating.

» Finally, don't let your dog become a pawn in your troubled relationship. Dogs can suffer from stress-related disorders as a result of constant emotional conflict in the home. Divorce if necessary; after all, a good dog is hard to find!

....

There are occasions when, for all kinds of reasons, behavioural modification doesn't have the desired effect for either the dog or the family concerned. Despite asking for several years, I still haven't been given a functional magic wand for Christmas, and the reality is that all my cases require hard work and commitment on behalf of the owners in order to change their dog's behaviour. In some situations, dogs are simply too entrenched in a habitual behaviour to be able to change. In others, the humans involved find it too difficult to alter their own habits, even once they realise that changing their own behaviour will have the greatest impact on the dog. Occasionally, there are emotional pay-offs for people in keeping their pets the way they are – as was the case with Sharon. These can be tricky for the behaviour specialist to spot, and even trickier to deal with.

Some years ago, I took the first-ever Master of Science postgraduate degree in Animal Behaviour to be run by a British university. Although I had already been practising for many years and ran a successful consultancy, I love learning and could think of no better excuse than to be able to research my favourite subject in the name of a new qualification. Part of my final dissertation examined the direct impact of behavioural help: how quickly any effects occurred and how long-lasting they were. My results were fascinating.

In the huge majority of cases I examined, a large part of the 'success' of the behaviour modification programme came from the fact that the dogs' owners learned more about the way their pets' minds worked. They came to appreciate that problem behaviour was a result of many different factors, and this understanding in turn facilitated acceptance. Not all behaviours can be changed in dogs, any more than they can in humans, and this is especially true if the dog came into the world with a predisposition to act in a certain way. Dogs are born with hard-wired drives and needs. If we believe that accepting this is important for dogs and humans to get along, then assessing a puppy is a skill that must come first.

JVK

6

Sociability

Choosing the perfect dog for you

Case history: **Jack the lad**

In my experience, those who read dog books are already hooked. They have a dog, or dogs, at home and are keen to learn more about them. It's ironic, then, that most dog books start with an optimistically titled chapter such as 'Choosing your dog' because, frankly, for most of us it's way too late.

However, there is still value in knowing how to choose a dog and what to look for – not in the traditional way of choosing between, say, a Cocker Spaniel and a Labrador, but with an eye to getting a dog that is going to be fun, good with your kids, a pleasure to be around and, most importantly, safe.

Dogs walk around with a full set of weapons in their mouths. The fact that they so rarely use them with intent to do real damage to humans is a miracle. However, the sad fact is that dog bites are all too common – ask anyone who works in an A&E department – and damage is most frequently done to people, especially children, by dogs that are known to the victim.

Why do dogs bite people? The answer is pretty simple: because humans annoy the hell out of them! From morning to night, people irritate dogs. Of course, we irritate each other as well, but we tend to get good at alleviating it by shouting or leaving Post-it notes declaring that we have left home for good or that the dinner is relegated to the bin. Dogs can't do this. They do try to warn us, in numerous little ways, when they get riled. If they were able to write, I'm sure most dogs would have used up a whole block of Post-it notes before lunchtime on the average day.

You think you don't annoy your dog? Did you eat something today that your dog didn't have? In my house, if my husband were to make himself so much as a cup of tea and not offer me one, there would be trouble, but there we sit, eating packets of crisps, biscuits and other delights, usually on the floor next to the dog's ear, in the car next to his head, or on the sofa close to his nose. I want to be clear here: I'm not suggesting that we should feed dogs our food. This leads to all kinds of behavioural and physical problems. No, if anything I would rather that dogs learn the clear and unfettered rule that they *never* get to eat human food, if only because this will help to reduce their otherwise almost continuous frustration about just how often humans eat without them.

It's not only food frustration that must cause irritation to the average hound. Every day, I cuddle my dogs when they probably don't want to be cuddled; usher them into the garden when they had been sleeping peacefully on the sofa only seconds before; move them from room to room, just by getting up and going myself, and then come back again to the first spot mere minutes later. I get ready to take them out for a walk, then make them wait while I talk on the phone. I look inside their ears. I pull their lips back. I stroke them when they don't want to be stroked. I tickle their feet. I pull them away from things they mustn't touch. I talk too loudly right next to their ears. All this, and that's without taking them to the vet, training them, or any of the other events that are probably all highly frustrating or purely nonsensical to dogs but perfectly rational to their human guardians.

So, perhaps the question is not 'Why do dogs bite people?' but 'Why don't they?' This can be summed up in one word: sociability. A truly sociable dog will put up with all the annoyances of everyday life with a human family because ultimately he is bonded, attached and dependent on you. True sociability is a wonderful thing. It is the buffer against aggression, allowing dogs not just to cope with humans and their odd ways, but to love them too.

For anyone who is thinking about getting a puppy or a dog – or if, like me, you know that your life will always be filled with dogs and you'd like to know how to pick the next one – here is the tale of one dog, chosen from a rescue centre: not for looks, not for his breed or type, but for his temperament.

Lucky, a small, short-coated crossbreed, had been in the rescue centre for four weeks. He was only six months old – a common age for dogs to be given up for adoption, simply because they are no longer cute puppies but have turned into wild and unruly adolescents. This happens remarkably early in dogs. It starts at around eighteen weeks and can last until the dog is eighteen months or so. Just in case you didn't know, wild and unruly adolescent canines are not dissimilar to wild and unruly human teenagers. They think they can fight the world, they always know best, they want to go their own way, they are obsessed by food and sex, they are lazy and clumsy. Oh, and they still need their mums. OK, dogs don't actually need their true mothers. Any sensible brood bitch starts to give clear messages to her offspring as early as seven or eight weeks that they should no longer be hanging around the nest. Nature is generally insistent that siblings and family members be dispersed before there is a risk of inter-breeding. However, adolescent dogs, just like adolescent kids, need security, guidance, help and support if they are to feel safe in the big wide world.

Lucky fitted this description to a T. He was only a little dog but his previous owners had circled the option marked 'cannot cope' as their reason for giving him up. Now, I must be clear: this is not a judgement. Neither does it mean that they didn't love the dog and want the best for him. In fact, giving up a dog is a heart-wrenching and horrible ordeal, and many owners who have to go through this do so only with the best interests of the dog at heart.

Lucky had come from a home with three small children, and although it's possible to raise kids and dogs together as a harmonious if slightly rowdy crowd, to be honest, it's tough. Lucky was one of those dogs who just wanted to be part of the fun. Bright and active, with a strong working drive from his part-Jack

Russell parentage, he couldn't help picking up the kids' toys, running off with them and darting under the furniture to play 'keep away'. However, it was his favourite pastime, chewing, that eventually landed him in the rescue centre looking for a new home.

Why would anyone want to take on a dog like this – or any dog from a rescue centre? First and foremost, it's important to emphasise that many, many dogs end up looking for a new home through no fault of their own. Dogs are often the innocent victims of marital problems, family break-down, household eviction, financial crises and home repossessions, as well as unwittingly causing childhood allergies or even – perhaps the saddest of all – finding themselves alone after the death of a loving owner. Such dogs are often house-trained, loving, easy-going family pets and they are just waiting for you. Others, such as Lucky, are simply being dogs. All they need is training and some healthy boundaries and, like most teenagers, they'll turn out just fine.

However, the dilemma for most prospective owners is that they don't know whether they will get one of these great dogs from a re-homing centre, or whether they will end up being the unlucky recipient of a dog that was placed in the rescue centre because of its behaviour. Sadly, this risk often makes the average family decide that it is somehow a safer bet to get a puppy from a breeder. This sounds alluring, simply because it is possible to predict that a puppy from two pedigree parents will look the same as them and (theoretically) will act the same too. However, what this ideal overlooks is that choosing a puppy, which will look and act more or less like every other puppy in the litter at the age of seven or eight weeks, is far more of a lottery than choosing a seven-month-old dog from a shelter, whose character and physical appearance is on show for all to see if you know what to look for.

Thankfully, learning how to pick a pet dog with a great temperament is relatively easy, and it's something anyone can do, simply because as humans we already know what signs of sociability look like. If you bear in mind that sociability is the prime characteristic to choose in a dog (not looks or because you feel sorry for them) then you won't go far wrong. Clear signs of friendliness in dogs look remarkably similar to those in people. Genuinely friendly people crinkle up their eyes when they smile. Their faces look relaxed and rounded rather than tense and hard. Their eye contact is soft and gentle. We move slowly and shake hands to show that we are not carrying weapons; dogs wiggle their bodies and turn so that they present their sides or flanks for petting, rather than presenting us with full-on teeth. This is important. To the unaware, scary dogs may look or sound as though they are simply over-excited, but they show their teeth from the front, stare directly at your face and approach you head-on. Imagine if a person you didn't know approached you like this. Would you feel comfortable? While many of the communication systems dogs use are different from ours, there are some fundamentals that are the same – and our gut instincts tend to be the best way of recognising when another being is acting in a defensive or threatening way.

Sarah first saw Lucky through the kennel bars. He was right at the front of the kennel, gazing at people as they passed by, and pressing himself against the wire in order to try and make social contact with them. Each time someone stopped to say hello, Lucky would squirm with wiggling ecstasy. His tiny body wagged all over and his tail went round and round in a windmill-style celebration of friendliness. The little dog's dark-rimmed eyes squinted up at folk as they poked their fingers through the bars and talked in baby voices, and his ears turned practically inside-out in response. Once the people had passed on, he sat and gazed after them, sideways to the bars, whining quietly.

After watching him for a while, Sarah approached Lucky's kennel. She crouched down and put her flattened palm to the bars. Lucky immediately came to sniff and lick her hand, and followed it when she moved her hand to a new position. At only six months old he was highly active but not manic, and although he was desperate to meet her, his facial expression was soft and gentle. Sarah asked him to sit. He looked at her quizzically, clearly pondering this strange request, then came back to be as close as possible to where she was crouching and pressed his side against the bars, exposing as large an area as he could to her touch. It was love at first sight.

Sarah asked the staff if she could get Lucky out of the kennel. At this point, most prospective owners are told to take the dog for a walk so they can get to know each other, but this does little to create a bond between human and dog. Most dogs are simply too distracted by the smells, sights and sounds of the great outdoors to be able to focus on anything else. Instead, Sarah and Lucky went into an indoor area and she watched him closely. After all, first impressions count.

At this stage, any prospective owner would be sensible to follow a basic strategy to find out whether the dog is right for them. Stroking the dog and admiring his or her good looks are insufficient. Instead, looking critically (though not negatively) at what the dog is actually doing is important.

Bear in mind a dog that has been in a strange kennel for a few days is effectively on his own. No longer part of a team, he is without the security of a familiar family, home, sights, sounds and smells. For most dogs, this is a highly uncomfortable and stressful experience and they are driven to align themselves as quickly as possible with friendly humans in order to feel emotional comfort and physical safety. The few dogs who don't care about this are the equivalent of independent loners – and while these dogs may be suitable as a working dog, they are certainly not for the average family home.

Sue Sternberg, a US shelter dog expert, has developed an easy, quick and remarkably reliable procedure for assessing a dog's sociability in just a few minutes. It's not definitive, of course, and can never give the true picture of the whole temperament and behaviour, but it does give the average person a head start in assessing whether or not this is the dog for them.

1. With the dog on a lead, stand still for about a minute and watch him carefully, but without interacting.

In this scenario, provided the environment is relatively free from distractions, the dog should want to make contact with you. Does he nudge you, look up at you, wiggle his body and otherwise attempt to elicit attention in a polite fashion? On the whole, dogs that have been

starved of human companionship by the confines of a kennel are really keen to count you as a friend, and they will make repeated attempts to get you to look at them, talk to them and touch them. Dogs that spend the whole of the minute at the other end of the lead looking into the middle distance and pretending you don't exist are telling you something: they don't want to interact with you.

2. Sit down and watch what the dog does now.
Sitting down makes you more accessible, and more available to most dogs. Continue to ignore the dog to give you a greater sense of the efforts he will make to create a

relationship between you. Bruisers may try to leap on you and test your strength by pushing with paws and mouth. Polite dogs will try repeatedly to make social contact, but in a way that feels more like asking than demanding. Dogs that stand apart from you, ignore you, or grab the lead with their teeth and bite or shake it are demonstrating that they would like to get away from you.

3. Stroke the dog from the back of his neck to the base of his tail, just once, then stop and watch. Dogs that are truly social love to be touched. Once they've got you engaged in physical contact, they will try to get more by turning towards you in a non-threatening

way, sidling up and making friendly facial expressions. If the dog moves away straight after you have stroked him, don't make excuses for him. Think how it would make you feel if someone you touched moved away immediately. If the dog goes completely still after you've stroked him or stares at you, put him back in the kennel immediately. These are both low-level signals of impending aggression.

4. Assuming that the dog has tried to initiate and maintain friendly responses, you can then pet and praise the dog for a few minutes, watching all the time what he does and how he responds. This is

different from thinking about how good *you* are feeling while petting the dog.

Different dogs have different responses to full-on human attention. Some get wound up and over-excited almost instantaneously. Others become calmer. Dogs that become more active when touched can be taught to be calm when handled, so this doesn't mean they should be discounted, but it's useful information which may influence your decision – especially if you have kids at home. Clearly, dogs that become calmer when touched by a human are easier to handle and less likely to become over-aroused and out of control.

....

Of course, as well as a rational analysis of the behaviour of a prospective new pet, there has to be a spark that kickstarts the relationship then makes it work, day in, day out for the next twelve to fifteen years. This chemical attraction – just like falling in love – is an integral part of the ideal choice, as it will buffer you against the bad times as well as enhancing the good.

Lucky lapped up the attention from Sarah, jumped on her knee when she sat down, and snuggled into the crook of her arm. He tipped up his face to gaze into hers, wiggled and made little squeaks of pleasure when she spoke to him. When she put him back down on the floor and stroked him, he deliberately moved towards her and stood calmly – quite a feat of self-control from a wired adolescent. The chemistry was all there.

It turned out that Lucky was half-Jack Russell, half-Chihuahua. His diminutive size was purely external because his mischievous personality made

him a big dog in a small frame. He's irrepressibly cheerful, tolerant, friendly and funny.

Owners of really great re-homed dogs don't think their dog was fortunate to find them and be given a good home; they thank their lucky stars every day that they got to take home the best dog in the world.

How do I know this? Because Lucky – now named Jackson – came home with me, was immediately adored by my whole family, and is happily snoozing on my knee as I write.

TOP TIPS FOR CHOOSING A PERFECT PET – PUPPY OR RESCUE

» **Physical features to consider:**

Size: think about the dog's strength as well as size. Strength equals risk where the dog lacks impulse control.

Coat length: grooming takes time and not everyone likes vacuuming every day.

Breed tendencies: if you are considering a crossbreed, think about how both parents may influence its behaviour. For example, I have a Collie/ Jack Russell-cross who is fast, lively and likes to bark while rounding things up.

Age: pups are fun but the time and training they need is intense. Older or even elderly dogs can be hugely rewarding and fit into a daily routine easily.

Don't choose a dog just because it looks cute or reminds you of a dog you used to know. Choosing a dog is like choosing a partner: select on looks alone and you'll be quickly disappointed!

» **Temperament features to consider:**

Positive signs of good temperament and sociability are easy to confuse with over-excitement and arousal. Tail wagging does not always mean friendliness; barking does not always constitute a threat.

» **Good signs**

- ✔ When in the kennel, the dog wants to approach you and make contact by pressing its muzzle or body against the bars. If you move your hand, the dog follows and tries to maintain friendly contact.
- ✔ The dog's body looks soft and wiggly.

- Facial expression is 'soft'. Eye contact is often squinty.
- The tail shows how a dog is feeling: low, quick wags may indicate uncertainty; tails that go round and round are usually showing pleasure and excitement.
- Once out of the kennel, does the dog give you attention? Social dogs want to solicit your attention: they will lean on you, nudge you gently and show affection, not just excitement.

» **Avoid**

- ✕ Dogs that stand still and bark at you from a little way back in the kennel.
- ✕ Dogs with rigid or tense body language.
- ✕ Dogs that show you their teeth, or stare at you with direct eye contact.
- ✕ Adult dogs that mouth you when they come out of the kennel – even if it's 'only in play'.
- ✕ Dogs that ignore you. Don't be fooled by the idea that the dog has been cooped up in a kennel and

therefore finds everything else distracting. If he's sociable, he'll be thrilled to have a human to give him attention. If you stroke him, he should ask for more, not move away.

How we choose our dogs is a complex and subtle process. I give lectures around the world, and I often ask the same question of attendees: 'If you were a dog, what kind of a dog would you be and why?' The answers are usually hilarious, as well as fascinating. Not only do people tend to identify very strongly with a certain breed or type of dog, but they will also tell me specifically what colour they would be, and whether they would be long- or short-haired! Some individuals claim to be like Labradors because they are good-natured, love food and are enthusiastic about everything. Others tell me they are like Wire-haired Fox Terriers: always on the look-out for action, fast-moving, and they occasionally enjoy a spat! This resonance doesn't seem to be based on the old adage that people look like their dogs of choice; it's more to do with the person's understanding of that breed's characteristics and behaviour. When I then ask them whether they chose the type of dog that they actually own, about two-thirds of the audience will give me an emphatic 'Yes'.

Do we choose dogs because we are trying vicariously to project on them characteristics that we wish we possessed, or because they *are* like us? If it is the latter, can we rapidly recognise similar characteristics and feel drawn to them in dogs, as we do in other people? I can certainly attest to this. I am lucky enough to own a soul-mate dog, a heartbeat at my feet, a dog of a lifetime. Tao is a Border Collie crossed with a Jack Russell Terrier. She's not everybody's idea of a perfect pet. She's fast, she's busy, she's determined. She's me! Seeing her in the rescue centre for the first time, I knew with certainty what she needed, and that it would be easy for me to meet those needs to keep her happy and healthy. I knew that we were meant to be together, and I knew that we would be inseparable by the end of the week. Fanciful? Perhaps, but it turned out to be true. This is the very chemistry of attraction – and it's a powerful thing at work.

Part 3

Health

7
You are what you eat

The effects of diet on behaviour

Case history: **Saffi, the irritable Golden Retriever**

Studies on the diets of children and young adults have shown that the food consumed is inextricably linked to the ability to concentrate and learn. Most parents know that some foods seem to suit their kids while others have a detrimental impact on their general behaviour. However, how often do we question the behavioural impact of what we feed our dogs? Trying to establish a direct causal relationship between the food they eat and their resulting behaviour is almost impossible in practical terms, simply because we can't isolate all the other factors that may be having an influence on the way they behave. However, preliminary studies do seem to suggest that food may have an impact on a dog's ability to learn, concentrate and solve problems. Add to this the huge amount of anecdotal evidence from owners who feel certain that their dog's behaviour is influenced by their diet, and it is hard to ignore.

Saffi was a beautiful, if rather restless Golden Retriever. At only two years old, she was at the peak of her fitness and her lovely coat shone in the stream of sunlight coming in through my consulting room window. Finding herself ignored, she wandered about the room, sniffing in the corners and generally amusing herself.

Saffi's owner, a professional lady in her early thirties, sat opposite me, gently rocking the back of a pushchair in which her fourteen-month-old daughter Lucy was fast asleep.

'I do hope you can help,' Caroline said quietly. 'Saffi is a wonderful dog, and until Lucy came along, she was our baby.' She looked a bit sheepish. 'I suppose that's why we're having problems. I'm terrified you'll tell us that she has to go.'

Saffi looked at me with big Golden Retriever eyes. How could anyone resist?

'Tell me what's been happening,' I said.

Saffi had been a lovely puppy. Full of fun and bounce, she was like any other Golden Retriever at that age. Her breeder had been sensible, careful and responsible – indeed, she vetted the couple who were so keen to take her home and even asked them if they were thinking of having children. They were honest and said yes, in the next couple of years, and the breeder told them how good Goldies normally are with children.

The couple had spent good time and money raising Saffi in the right way. They had attended puppy classes and enjoyed them, even winning a rosette on the final evening. They had invested in Saffi's training and had taken her on a good basic course. She would sit, lie down and come quickly when called. Her house manners were excellent and apart from being somewhat over-enthusiastic when greeting visitors on occasion, she loved everybody and anybody.

When Caroline became pregnant, the couple had no undue concerns. Of course, they were aware that Saffi's routine would change a little and that she might not get the same amount of attention and exercise she had before, but they were sure she would be fine with the new baby and that she would accept her as one of the family. They even had a private chuckle about a rather over-zealous health visitor who told them Saffi should be kept permanently kennelled outside and not allowed near the baby.

Caroline swallowed hard as she recounted what happened next. She had been warned that she might need to stay in hospital for a couple of days after the birth, so they had sensibly booked Saffi into a local boarding kennels for the week. She had been there before, so her owners rationalised that it was better for her to be looked after like this rather than left alone for long periods. Once home from hospital with baby Lucy, and after a couple of nights to settle her in, they picked up a rather bouncy Saffi and the whole family tried to establish a comfortable new routine.

The couple had read up on how to introduce dog and baby safely, and this appeared to go well, with Saffi sniffing at the baby's clothing and then running to get her a toy. Overall, Saffi seemed interested in the baby, but was more inclined to leave the room if she was crying than want to interfere in any way. In retrospect she perhaps seemed a little more restless and agitated than previously, and she barked more, but they put this down to reduced exercise and a lack of attention – almost inevitable in the first few weeks when they had their hands full.

However, as time went on, Saffi seemed to become more and more irritable. She would slink off into the garden at every opportunity and would sneakily eat grass, soil, even stones from the edge of the flower

borders. If called to come indoors again, she would ignore her owners completely, and she would run away from them if they came out to get her: infuriating when you are holding a baby under one arm at the same time. In the house, she seemed like a different dog from the one they had known and loved. Caroline described her as having 'sullen moods' – which might seem far-fetched when describing a dog, but actually gives an accurate description of an altered emotional state. She tried to avoid being petted and stroked, and was beginning to shy away from any contact with her immediate family.

Husband Tim was becoming increasingly concerned about the dog's behaviour. He called a local dog trainer, who said she thought it sounded as though Saffi was jealous of the baby and was trying to get back the 'pack equilibrium'. The trainer recommended putting a long line on the dog and pulling her back into the house if she refused to come when called. After all, dogs should always be made to respond if you have told them to do something, shouldn't they? The couple tried this a number of times, but anyone who has ever seen a Golden Retriever digging in her heels will know just how impossible this can be. Slumping to the ground and using her 'dead weight' against the pull of the line, Saffi wasn't going anywhere.

Over the next few weeks, the couple tried less confrontational methods to get Saffi to comply with their wishes. She would sometimes come to them for food, but if she was in a really bad mood, she would head out the back door, crawl behind a bush at the top of the garden and simply stay there by herself until she felt ready to come in. If they shut her indoors, she would whine and pace by the French

windows almost continuously until they let her out again.

All this was distressing enough, but her attitude towards the baby was also beginning to change. 'At the outset, she just seemed uninterested,' said Caroline. 'Now, she'll do almost anything to get away from Lucy. And although Tim says it's silly, I don't like the way she looks at her.'

Right on cue, Lucy stirred in her sleep, burbling and moving her hands and arms. Saffi became still on the other side of the consulting room, where she had been rooting through the waste-paper bin. She glanced anxiously towards the baby and ever so slightly rolled her eyes.

Now, humans and dogs may speak a very different language, but we do share one common skill. As social mammals with highly developed communication systems, we are experts at detecting tiny signs and signals that mean trouble is brewing. This aptitude is so sophisticated that very often we can't even describe what it is that we saw or heard or felt – we just knew it predicted some kind of threat. We call it intuition, or gut feeling, but really we are reading the tiny signals of intent that others unconsciously give out, whether human or canine. In his book *A Gift of*

Fear, Gavin de Becker discusses how survivors of violence – or those who manage to avoid violent attacks completely – will often explain that they 'just knew' to act in a way that kept them as safe as possible. This is not some kind of sixth sense but an incredible ability to pick up on minute details that don't fit with our usual picture of safe events. In one of de Becker's fabulous examples, he explains that the woman who chooses not to get into a lift with a stranger, despite the slight social awkwardness that it causes, probably does so not because she is paranoid or generally fearful, but because she unconsciously noticed subtle but tell-tale signs that might mean trouble. Maybe the man looked at her a fraction too long, maybe he unconsciously made a tiny forward movement rather than stepping aside, or maybe another woman getting out of the same lift looked slightly flustered. Responding appropriately to such signals can help to keep us safe, just so long as we prevent our 'rational' brain from interfering and convincing us that everything is alright.

Instinctively, humans are pretty good at reading tiny body language signals or facial expressions in dogs, just as they are in other humans. It's just that they generally can't tell me *what* it was they didn't like.

I have come to truly appreciate such knowledge. Not only does my ability to read these subtle signals help to keep me safe when working with dogs, but it also gives me wonderfully useful information when passed on by an observant and sensitive owner.

In Saffi's case, Caroline was picking up on some pretty important indicators. Contrary to popular belief, dogs do not growl or show their teeth as a first line of attack or defence. Instead, just like humans under threat, they go still. This is called the freeze response. It may only last for a split second, or it may completely immobilise the person or animal for minutes or even longer. Indeed, survivors of great trauma sometimes report that they simply couldn't move when first faced with the terrible incident – and that others simply couldn't snap out of their helpless, frozen state of fear at all. Such immobility is one of the remnants of the 'caveman survival kit'. When faced with a sabre-tooth tiger, it might have been a more sensible policy to freeze and hope you weren't noticed than it would be to try and run away. However, in the event that the threat continues to approach, it should be very much the first response and not the last.

Saffi's slight eye rolling was another indicator of arousal and uncertainty. This is most often seen when dogs are defending something they have within their possession. A typical posture will show the dog with his head over the item, still, and with the white of the eye showing as he rolls his eyes to look at the perceived threat.

Saffi's freeze response came as a direct result of Lucy's activity. It therefore figured that Saffi's anxiety about Lucy would get worse as she became more mobile and, when I asked Caroline said her feelings of disquiet around her dog and baby had increased since Lucy had begun to crawl.

Saffi came and sat by my leg. She glanced frequently at the pushchair then away again, checking out the source of the threat. I stroked her silky head, and – as she was leaning against me and giving body signals that meant I felt completely safe – ran my hand down her back. Her skin rippled as if touched by an electric current. She moved away from me immediately, sat down and scratched. Caroline, who had been attending to the baby, hadn't seen this. I asked her whether she had noticed Saffi's skin being sensitive to the touch or whether she had seemed itchy in the last few months.

Yes, she said, she had noticed the skin crawling. In fact, she had mentioned it to the vet on her last visit when Saffi went for her routine booster vaccinations. The vet had said that it would be sensible to treat Saffi for fleas (which they had) but the odd symptom seemed to have remained.

'To be honest,' Caroline said, 'she doesn't seem to want to be touched at all, and she used to be so cuddly. Now all she seems to want to do is sit under that bush and eat rubbish.' Her sigh said it all. 'She's just not the family dog she used to be.'

Dogs are good at communicating with humans; so good that we sometimes forget they are a completely different species, without the ability to tell us explicitly what is going on in their heads. This may seem obvious, but have you ever considered that your dog might have the occasional headache? How would we know if they have stomach-ache or are feeling off colour? To be honest, by the time the average human notices that a dog isn't eating, doesn't want to move about or just 'isn't his usual self', the symptoms are pretty well advanced. This is particularly the case with dogs that are what I describe as 'stoical'. Out of my three dogs, I have one that will scream if you so much as look as though you might clip her toenails, let alone give her an injection or administer ear drops. Clearly, she is a sensitive type, especially to touch. However, I have another dog – a Golden Retriever – that doesn't even look round if you stand on his tail. Even when in pain, he doesn't show it – and this stoicism seems to be a Goldie trait.

There are many clinical causes of behavioural problems, and in stoical or adrenaline-charged dogs these can be particularly difficult to diagnose. For this reason alone it is important that all dogs showing unusual behaviour are seen by a behaviour specialist only on referral from a qualified vet. In my practice, we work alongside vets and often need a dual approach combining their clinical expertise and our behavioural understanding in order to get to the root of a problem and resolve it.

There's a little-known clinical condition in dogs – and more commonly cats – called tactile hyperesthesia. This is a disorder that causes the skin to crawl and ripple uncontrollably, making the unfortunate animal leap up and attempt to bite or scratch the area in order to gain some relief. In a few cases, the irritation can be so severe that the cat or dog damages the skin with its teeth or nails and effectively ends up self-harming in order to try and bring some relief to the random twitching. While some of these cases can have an underlying neurological basis, what causes the condition is not fully understood. There are a number of possible factors which might have an influence, and other clues in Saffi's case were leading me to suspect what might be behind it.

During the rest of the consultation, I asked Caroline about Saffi's routine. We talked about her strange obsession with eating unusual items, a behaviour called pica. I already knew that she had a taste for soil, grass and stones, and wasn't at all surprised when Caroline told me that she also had an obsession with tissues, both used and unused. Apparently, she had also had a penchant for eating coal, until her owners had blocked her access to the coal hod in the garden shed. I asked what impact these 'extras' seemed to have on Saffi's digestive system: although she seemed perfectly healthy she did seem to need to go to the toilet a lot – but perhaps this wasn't a surprise – after all, what goes in must somehow come out.

Finally, I asked Caroline my jackpot question. Had Saffi only been this way since she stayed in kennels? Perhaps the changes the family had seen were not directly linked with the baby at all but with the fact that … I took a breath, but it was Caroline who finished my sentence.

'When she was in kennels they changed her diet because she wouldn't eat. When she came out we decided to stick with the one they had used because she seemed to like it so much.'

Bingo.

Dog food in this country, and elsewhere, is a multi-million pound industry. Thanks to clever advertising, we all know about the major brands, the colours and the shape of the packaging we are looking for in the supermarket, and yet the contents of those tins and packets remain a bit of a mystery. Do you know what is in the food you are giving to your dog? Is the fact that the dog likes the food enough? Are you happy dishing up something that might look appealing to you but is really made from the beaks, claws and feathers of chicken carcasses? A closer inspection of the contents of some dog foods can be rather a shock. Did you know, for example, that a food could claim to be 25 per cent protein yet be made of old shoe leather and sawdust? OK, this is an extreme example, but the message is that in order to choose a dog food that's right for your pet, you need to look at the ingredients list not just the percentage analysis.

The ingredients of the food must be listed in order of quantity. This means a food that lists turkey meat as the number one ingredient has more turkey meat in it than anything else. However, if it lists something vague you don't recognise, such as 'meat and animal derivatives', then you need to ask yourself what these are. In fact, they can include heads, feet, guts, lungs, feathers and even wool. While these may do no harm to your dog, their nutritional value is questionable as they are hard to digest. Many dog foods also rely on high quantities of undefined cereals. These are cheap and although they act as 'bulking agents', again their lack of digestibility makes them of little nutritional value.

Research throws some interesting light on the impact of diet on canine mood and behaviour – and this is not just about what we feed, but how we feed it too. Many people will confirm that when they are hungry low blood-sugar levels can make them lethargic, slow and sometimes downright irritable. From much anecdotal evidence, I would put money on the fact that dogs experience this too – especially if they are only fed one meal a day and have to wait 24 hours for the next. Frankly, I'd be pretty grumpy too.

Saffi's owner hardly needed convincing to try a change of diet for her dog. After all, it's probably one of the least onerous of all the interventions we can make and it usually only takes two to three weeks before behavioural changes are noticed. However, I could tell that although Caroline was happy to try it, she didn't really believe such a small change would be enough. Saffi lay quietly on the floor, keeping one wary eye open for signs that Lucy might head towards her. We chatted through general management options to ensure Lucy's safety, and promised to speak again in a fortnight.

To my surprise, I didn't have to wait that long to hear from Caroline. After only four days she called, and I could hear Lucy gurgling in the background.

'We can't believe it,' she said. 'The difference is amazing. Saffi keeps coming into the house. She wants to sit next to me, and didn't even flinch when Lucy put her hand on her neck. I know it's not a miracle cure and we still have some way to go, but after such a short time, all I can say is that we've got our dog back.'

Clearly, such extreme improvements are rare but they can – and do – happen from time to time. Sometimes in dog behaviour, as in so many other areas of life, it is not the amount of leverage that is applied but knowing where to apply it that really makes the difference.

TOP TIPS ON FEEDING FOR OPTIMUM HEALTH AND BEHAVIOUR

- Diet can affect canine behaviour just as it can our own. If your dog has two or more of the following habits, then a diet change may be useful:
  - Eating unusual items, such as tissues, sticks, grass or soil
  - Showing over-active behaviour and a lack of concentration
  - Inconsistent digestion and large, smelly faeces
  - Itchy skin (when there are definitely no fleas or other creepy crawly infestations)
  - Hard biting, over-excitement and 'tantrums'.
  - Consult a vet if you are unsure.

- To introduce a new diet, replace the old food with the new bit by bit. Don't change the diet suddenly or it might trigger a tummy upset.

- The ingredients of the food you choose are far more important than the percentage analyses shown on the packet. Their quality (and how digestible they are) are also more important than how the food looks, or what the packaging is like.

» If you are feeding a good-quality complete food, don't be tempted to add other foods to it. This can effectively unbalance the food and can lead to behavioural issues.

» Feeding twice a day helps to balance your dog's blood-sugar levels. Like us, dogs can get irritable if they are hungry.

Owners often ask me which is the best food for their dog. This is a million-dollar question that's perhaps best directed at your vet. There are as many answers as there are dogs on the planet, because each dog will have his or her own needs and requirements. Just as humans can be intolerant to a particular food or ingredient, so can dogs. The difference is that we can verbalise our discomfort and make decisions about whether or not we eat that food again.

What I would emphasise, however, is that dogs nearly always do better when the ingredients are good quality and as close to natural as possible. Dogs whose digestive systems have to work hard to extract nutrition from their food may suffer from lack of attention, concentration or impulse control – making them prime candidates for a misdiagnosis of dominance or disobedience. Correct feeding is also a case of human honesty. Just because a dog is a working breed doesn't mean that he needs to be fed a working-dog diet. Two walks a day don't constitute work!

On the whole, we recognise the effect of diet on ourselves and our families. We know when we've eaten too much, when we've been eating (or drinking) things we shouldn't, or when we haven't given our bodies and minds sufficient 'fuel' for the day. Once aware, owners can be just as savvy about the food they give their dogs. They can see how low blood sugar might make the dog feel irritable, or how simply altering the diet for a couple of weeks can have an impact on a re-training programme, but making this link in the first place can be quite a leap. Perhaps it is the lack of serious research in this area that is holding us back, although the difficulty in excluding other factors to isolate the effects of diet alone contributes to the dilemma. No matter the reason, it is certainly a hot topic of debate among practitioners in the field of animal behaviour, and it's one that undoubtedly deserves to attract more investigation.

8

Stress and anxiety

On the couch

Case history: **Zeus, the agoraphobic Afghan Hound**

So you think you've had a bad day? Often canine behaviour is labelled as 'naughty' or out of control when really the dog is ready to implode – or explode – with anxiety and stress. This is because in dogs, just as in us, the brain is designed to look for threats and to act upon them there and then. This is a magnificent survival strategy, and one most humans relate to but rarely consider to be a part of their dog's basic functioning.

Humans may come home from work tense and fatigued by the sheer hectic nature of modern life, but at least we can rationalise it. We talk about it endlessly with loved ones and plan our retirement as a bit of escapism. Dogs can't do this. For a dog stressed by the sheer presence of upsetting stimuli in the environment, life can be one long stretch of anxiety, punctuated by brief spells of respite or relief.

While certain parts of the dog's brain can become 'hijacked' by fear or stress, causing reflexive reactions, including aggression, other types of long-term stress can cause illness, depression and even obsessive-compulsive disorders, such as self-harming – just as it can in humans. In my practice, I have seen dogs that have chewed off their own tails or made terrible open wounds in their limbs from continuous nibbling, and there was one dog who even had to have the tip of his penis surgically removed after sucking it relentlessly as a response to anxiety.

Dogs may experience stress or anxiety for a number of different reasons. It may be that, like Amber the Cocker Spaniel in Chapter 3, they simply didn't experience enough in their early weeks to build coping strategies and realise that life just happens. For others, it can be that their environment has changed – and they have not. Sadly, this type of anxiety is sometimes seen in dogs that have been imported from other countries, for whom their new home must seem to be on a different planet from the one they left. For a few individuals, a single bad experience may be enough to convince them that life is continuously threatening, while for others it may be a succession of incidents that convinces them to be on high alert.

How dogs cope with different situations is hugely influenced by type, breed, genetics, previous experience and even their mother. Studies have shown that puppies born to mothers of a nervous disposition are – surprise, surprise – more likely to be nervous themselves. While this might seem obvious, what may not be so apparent is that puppies born to mothers who are stressed while pregnant are also more likely to be anxious dogs as adults themselves – proving that the physiological effects of stress go far deeper than the eye can see.

Zeus was referred to me by his owner's vet as a 'lost cause'. Apparently, the owner had sought advice from several dog trainers about Zeus's escapology from the garden, all to no avail. Now, as any professional will tell you, this is usually a bad start – not because the dog is beyond hope, but because the owner may feel that she has already tried everything. This can lead to feelings of despondency – or worse, of hope that my magic wand will somehow be more effective than anyone else's! I therefore called Jenny with a big dose of trepidation.

'Your vet says you would like to have a chat about Zeus,' I started, cautiously.

'Ah yes,' said a well-to-do voice on the other end of the line, 'we're at the end of our tether, you know.'

I asked her to describe the problem.

'Well, he just won't get out of bed.'

This wasn't what I had been expecting. 'OK,' I said, rather stumped. 'I was led to believe that the problem was something to do with escaping from the garden.'

'Oh, that was weeks ago. Now he won't get out of his bed, even for a walk. We just don't know what to do.'

As I do with many of my cases, I asked Jenny if it would be possible for her to send me a video of Zeus's behaviour. 'There isn't any,' she said. 'He doesn't behave, he just lies in his bed. But I'll send you a clip anyway if I can take it on my mobile phone.'

Bemused, I gave her my e-mail address and politely bade her goodbye. Fifteen minutes later, I was amazed to see a video file waiting in my inbox. The short clip showed a dog in its bed. It stared at the person taking the video with dark and sorrowful almond eyes, then hid its head inside the basket. With only this much to go on, I called the owner again and we arranged that I should make an urgent home visit.

I arrived at Zeus's home the following afternoon. Following my instructions to use the speakerphone at the entry gate, I waited to be buzzed through and drove slowly under a canopy of trees, across what can only be described as stunning open parkland and up towards a magnificent house. This was clearly an impressive estate and I could see how a beautiful Afghan Hound would fit right in, cantering across the manicured lawns with flowing locks flying out behind. I leaned against my car, admiring the view, and rather wishing I had worn a jacket over my shirt, when I realised what was missing. No barking. Not a sound, in fact. This is pretty unusual; in fact, it's so unusual for me not to be greeted by barks when I do a home visit that I suddenly panicked I was in the wrong place.

Jenny answered the door and welcomed me in. The house looked like something out of *Gone with the Wind*, with a huge sweeping staircase and a hall that would comfortably have housed a whole herd of Afghans. Jenny showed me into a drawing room. I sat down. We had tea. However, there was still no sign of Zeus.

Reluctant to comment on a dog I hadn't yet seen, I asked Jenny where he was and she showed me to the kitchen. There was a large basket in the corner of the room, containing an equally large, silky dog. He looked at me with big, brown eyes and flicked the end of his tail in greeting but didn't get up.

'And you've had him checked by the vet?' I asked.

'Oh yes, as soon as I get him into the car, he's completely fine. He bounced around the vet's surgery as if nothing was wrong,' Jenny answered. 'It's just in the house and garden he's like this.'

I produced some deliciously smelly pieces of hot dog sausage from my bag, and dropped a couple on the floor near Zeus's basket. His nose twitched in their direction, but it was clear that he wasn't going to move.

'Come on, Zeus,' said Jenny in an encouraging tone. She patted her knees, but Zeus gave a large and sorrowful sigh before tucking his head back between his front paws. He seemed to be trembling slightly and looked even more anxious.

We retired to the lounge to talk. Jenny told me that until three weeks ago Zeus had been an outgoing and exuberant dog. So exuberant, in fact, that his owners had trouble containing him in their huge gardens, as he would take any opportunity to get outdoors and dash off into the trees, disappearing after rabbits or squirrels or some other quarry, and deaf to their calls to return. Jenny explained that the whole estate covered around 40 acres. As it was unfenced, it meant that on several occasions Zeus had been missing for several hours and she had been beside herself with worry in case he had been hit by a car, had strayed beyond their boundary or had even been stolen.

Jenny's efforts to train Zeus to come when called had failed completely. In desperation, she had called Zeus's breeder for advice, to be told that Afghan Hounds were untrainable and that they should never be allowed off lead. Considering this impractical, Jenny employed a series of dog trainers who tried – and failed – to establish a reliable recall with a dog that had been enjoying the taste of freedom for several months.

'In the end, we decided it was simpler just to put a fence in,' said Jenny. 'In fact, my husband reckoned it was cheaper than employing trainers.'

'So you fenced the whole estate?' I asked, amazed.

'Oh, no. It's only the area around the formal gardens.' Jenny waved her hand at the window and I got up to have a look. I couldn't see a fence.

'Is it an electric fence? When did you have it installed?' I asked.

'It's an invisible fence,' she corrected me. 'We had it put in, oh, about a month ago. I bought it online. It worked brilliantly.'

Hmmm. It may have worked brilliantly from a human point of view ...

We went back to where Zeus was still lying in his bed. I asked Jenny to put his lead on and we managed to get him to stand up and walk towards us. As we headed out of the kitchen towards the hall and front door, he pulled in panic, clearly trying to escape some terrifying, formless demon. When we took him the other way, towards the door to the garden, the big dog simply froze, shook like a jelly and slumped to the floor in total helplessness. This had all the hallmarks of a genuine phobia. As the poor dog lay in a heap, I checked his collar to make sure it wasn't too tight and immediately felt the bulk of an electronic device under the hair on his neck.

'You never saw this behaviour before the electric fence was installed?'

'Invisible fence,' Jenny corrected me. 'No, but why would that upset him so much? The collar's not heavy and even if you take it off, it doesn't make any difference. Zeus still doesn't want to go out.'

Jenny removed the collar containing the receiver box for the electric fencing, and handed it to me to make the point. The collar is designed to emit a beep as the dog gets close to the perimeter wire, so that it learns not to stray beyond a certain area. If it doesn't stop, it will receive an electric shock. However, the unfortunate reality is that most dogs receive several shocks before they learn that the beep precedes the shock.

Anyone who has ever accidentally grabbed an electric fence (or, worse still, attempted to climb over one) will know that 'static impulses' – as electric shocks are sometimes called in the marketing – are highly aversive. If they weren't, they simply wouldn't persuade animals not to cross them. The use of electric fences for dog containment is controversial, and the bottom line is that when using any punishment in 'training' there are unforeseeable risks. These may be mild – a dog that is smacked for barking in the garden may simply learn to avoid his owner when out there – or they can be far more serious, such as the dog that is yanked on a choke chain for pulling on the lead, but associates the pain with seeing a young child across the street.

Zeus had been wearing the receiver collar for a month. His owners had followed the instructions on the box and Jenny reported that Zeus had received several shocks before learning that the beep sound on the collar meant he was too close to the fence line, and he had started to avoid it by running back to the house and the safety of the kitchen. She said he seemed reluctant to go into the garden after that, but they were not too concerned at first because it meant their problems with his escapology had ceased completely. Sadly, so had Zeus's *joie de vivre*.

Stress symptoms are frequently overlooked in dogs. All too often owners will report that their dogs are excited, panting and even jumping up at them, yet they interpret this as naughtiness, not stress. More introverted dogs may become quiet or passive, may

move slowly or low to the ground, and will hold their ears back and eyes wide as they try to hide, or seek reassurance from their owner.

Stress symptoms can vary widely. Few people would imagine that dogs can have sweaty paws, just as humans get sweaty palms, or that they sometimes shed hair when stressed or anxious. Understanding the different ways dogs can demonstrate stress should be high on any owner's list because it will put you in a position to recognise their emotional state and take steps to alleviate it. Your dog will thank you for it.

Symptoms of stress

- Avoiding eye contact
- Barking
- Loss of bowel control
- Destruction or chewing
- Digging to escape
- Dilated pupils
- Ears held back
- Extended or lolling tongue
- Gaping and panting
- Lip licking
- Loss of appetite
- Over-activity
- Pacing
- Reduced activity
- Salivating
- Scratching
- Shaking or trembling
- Shedding hair
- Sniffing the ground
- Sniffing or licking own genitals
- Sweaty paws
- Tail chasing
- Urination
- Uncertain tail wagging (low and fast)
- 'Whale eye' (white of eye revealed)
- Yawning

Still holding the collar in my hand, I wandered back down the passageway. Little Albert popped into my head. Little Albert? For those not acquainted with this unfortunate child, his story is a sad but fascinating one. In 1920, a baby boy of only nine months was chosen to take part in a psychological experiment at Johns Hopkins University in Baltimore. At the start

of the study, Little Albert was shown a white rat. He had no fear and tried to play with it. In subsequent trials, however, each time the baby touched the rat, the experimenters made a sudden noise by striking a steel bar behind his back, making him cry and show fear. After several pairings of the loud noise and the rat, Little Albert began to show fear every time the rat was brought into the room. While this is hardly surprising, what was interesting about the experiment was that this fear response quickly generalised to other 'furry' things, and very soon Little Albert was also showing fear of rabbits, dogs and even a seal-skin coat, despite never having been worried by these things before.

I stepped out into the garden and walked up to the fence boundary line, where the collar beeped in my hand, warning me that I was close to the wire. The setting was on the highest level and not exactly relishing the idea of receiving a shock myself, I walked away and back to the house. As I came in through the kitchen door, the beep went off again. Surprised, I retraced my steps. No sound came from the collar. I walked back again. Still nothing. Maybe the collar was faulty. I laid it on the table. A beep sounded again, but it did not come from the collar – it seemed to come from the wall behind me, right above Zeus's bed.

'What is that?' I asked Jenny.

'That noise?' she said. 'It's the gate entryphone. It beeps when the gate is open. There's another one in the lounge.'

Zeus hid in his bed, quivering from top to toe. As a smart dog, he had worked out pretty quickly that a beep meant – quite simply – that an electric shock was coming. If the sound had only ever occurred in the garden, he could have learned to avoid the area in which it happened, but here in the house – in his own bed, which was meant to be his sanctuary – he found himself unable to escape from the almost continual threat of random punishment. It didn't matter that the beep came from somewhere other than the collar because it was similar enough to have been paired in his mind with the aversive experience of being shocked before. Just like Little Albert, for poor Zeus there was no relief. Whatever he did, wherever he went, the warning beep still told him he was about to be subjected to a shock – and having run out of options, he simply tried not to move at all. Such a state is known by scientists as 'learned helplessness'. It is a complete withdrawal in the face of chronic or continuous stress or anxiety. The animal just gives up. For any creature, being subjected to an unpredictable and ongoing threat has serious mental and emotional implications. Indeed, in human society we describe such experiences as mental torture – and it is with good reason that such strategies are generally outlawed, even in warfare.

Zeus's anxiety at the prospect of being given random electric shocks was not simply going to disappear with the removal of the collar or the invisible fence. Even after the entryphone system was disabled so that it no longer beeped in the kitchen or the lounge, he crept around the house on his belly, preferring to hide in his bed or behind the sofa. This may sound odd, but just because the signal of threat is gone does not reassure the subconscious that all is well. Anyone who has ever had a car accident knows this. It doesn't matter that you survived, that you now drive a different car, that the circumstances are not the same: whenever you drive in that particular area, or across that particular junction, your subconscious reminds you that there might be a threat. It can be enough to get your heart racing, your palms sweaty and your mouth dry – all based on emotional memory.

Jenny and I worked for six weeks to try and help Zeus recover, but progress was painfully slow. At the point where the vet was considering using drugs to ease the dog's anxiety, Jenny's husband was told that he would be leaving the country to set up a new work project overseas. After much soul-searching, the couple decided to re-home Zeus rather than subject him to the trauma of travel and the disruption that comes with such major relocation. Zeus went to live with a lady in a tiny terraced house on the coast.

Despite the considerable downsizing of his abode, Zeus's problems literally disappeared overnight.

Once free of all the environmental reminders of his previous anxiety, he returned to his previously gregarious and energetic self. And all that escapology? It was put to good use as Zeus's new owner took him Afghan racing nearly every week, just for fun. Not only did he look breathtaking as he sped round the course after a dummy hare, but he turned out to be a winner too.

TOP TIPS FOR RECOGNISING AND EASING STRESS

» Learn to recognise how your dog expresses stress. Some dogs become more outgoing or appear to be naughty, while others become withdrawn.

» Be aware that dogs (like us) don't learn well when stressed. For this reason, training needs to be stress-free to ensure optimum learning.

» Some stress symptoms can have other meanings, so don't get hung up about it if your dog yawns, shakes or scratches once in a while. Look at his or her behaviour in context rather than leaping to the conclusion that he's stressed.

» Many dogs are stressed by loud noises, particularly fireworks and thunder. Giving him or her somewhere to hide – such as under a heavy blanket or in a small space covered with towels – during a storm or on Guy Fawkes Night can help. Body wraps can also be useful (see page 36).

» Puppies that are exposed to the sights, sounds and smells of the world around them when they are very young can build defences against anxiety. Such 'stress immunisation' is essential and needs to happen before they are sixteen weeks old.

» If your dog is stressed about something, it's kindest to remove him from the stressor. If this is impossible, try to minimise the stress by behaving in a confident and nonchalant way yourself.

» Bear in mind that dogs can be stressed about almost anything – just as we can. It doesn't matter if the stressor seems silly or trivial; if it's causing stress, your dog needs help.

» Anxiety is often caused by the dog pairing two unconnected events, especially unpleasant events with things they see or hear. Just think of Little Albert. For this reason, punishments and aversive-style training are risky and should be avoided.

Zeus lives in my memory as one of the saddest cases I have seen. Although the situation was resolved, the fact that he had to endure such extreme anxiety for even a few weeks and that the extent of his misery seemed to fly under his owners' radar was upsetting. Of course, to a certain extent, all animals – including us – have to learn to cope with day-to-day stressors, but we all have a limit after which stress becomes distress, and this will vary from individual to individual.

Tolerance to emotional stress will also depend on the nature and duration of the stressor. In human psychology, it has long been known that the more unpredictable and the less controllable a stressor is, the more impact it is likely to have. The now-famous experiments of Seligman and Maier in the 1970s demonstrated that rats subjected to prolonged exposure to unpredictable and uncontrollable stress developed long-term responses akin to those many humans would recognise. A 'normal' rat that had had no previous stressful experiences was placed in an area divided into two. A mild electric charge was passed through one half, preceded by a warning signal, and the rat quickly learned to jump to the 'safe' half of the area when the signal was given. Whilst unpleasant, the experiment showed that rats are more than capable of removing themselves from potential harm after they learn the associated signal. However – and here's the important part – when they used rats that had previously been given random electric shocks over a long period of time, with no warning and no escape route, even if given the chance to jump to safety, they didn't. Such inability to move away from the cause of pain, stress and misery was a direct result of the animal having experienced a loss of ability to control and predict, and this quickly became generalised so that it simply did nothing to help itself. In animals, we call this 'learned helplessness'. In humans, we call it depression.

Such extreme and acute psychological stress is thankfully rare in dogs, but when it does occur we must always look at both the cause of the stressor and the dog's own ability to cope. Where the stress is at a lower level but is more chronic, some of the same learned helplessness may occur. This can result in suppression of behaviour – in other words, the dog doesn't do very much at all. In cases where this follows a 'rank reduction' programme, confrontational training methods such as the alpha rollover, or 'teaching him who's the boss', we must be careful not to celebrate this lack of normal canine behaviour – this depression – as evidence that the so-called 'training' has worked.

9
Sex

Hormonal hounds

Case history: **Toddy, Trixie and Tilly – a hormonal hat-trick**

Neutering is a topic that can have strident women shouting and nervous men sweating! It seems to cause impassioned and dogmatic (excuse the pun) feelings, which are rarely based on clinical evidence. Blanket advocates of neutering tend to regard owners of unneutered dogs as irresponsible. Blanket anti-neutering folk regard it as unnecessary mutilation – and cross their legs when merely thinking about it.

Before I run the risk of getting hate mail on the subject, let me say this: I feel strongly that neutering needs to be considered on an individual basis every time. There is no one rule that can be applied to all dogs, no matter what gender. In the vast majority of cases, neutering prevents unwanted puppies, keeps dogs out of trouble and can help to prevent certain clinical diseases. However, it can also have its drawbacks and in a few cases, neutering is the very last thing the dog needs in order to be a well-balanced canine.

Toddy is a healthy, hearty Chocolate Labrador. It's little wonder the colour is described that way as his shiny rich brown coat looked for all the world like a slab of my favourite treat.

At thirteen months of age, Toddy sniffed round my office like any other adolescent: ignoring his owner and vacuuming up scent through his nostrils as if it were a drug. An affable dog, he didn't object to my presence nor did he elicit attention from me, unless you count sniffing avidly at my trousers. Toddy's owner was almost beside herself with anguish. This was her third Labrador, and none of the others had behaved this way. The problem was that although dog and owner enjoyed a good relationship, Toddy demonstrated remarkable and profound selective deafness whenever he was in the park, and most especially whenever he spied another dog – no matter how small a speck it was in the distance. This had led to some pretty hair-raising incidents, including one in which Toddy had shot off in search of a playmate and had crossed a main road in order to reach him.

I asked Toddy's owner, Elaine, exactly what she had tried so far to ensure that Toddy come back when she called. Looking pained, she pulled a piece of paper from her handbag and started to reel off a list that pretty much encompassed everything any trainer – good or bad – is likely to think of in order to improve a dog's responsiveness in the park. Elaine had used treats to reward Toddy if he did come when called. He still ran off. She kept him on a long line for several weeks but as soon as she let him off again, he was gone. She had divided his dinner into portions and taken them out on walks, but still he would rather play with another dog even if it meant going hungry. Looking embarrassed, Elaine also confessed to having used a remote 'spray collar', which squirts the dog with citronella or a jet of compressed air if it misbehaves at a distance, but she said this seemed to make him run even faster – because, of course, dogs soon learn that even the most unpleasant pieces of electronic equipment have a limited range and the faster they run, the quicker they can get beyond them.

To me, Toddy looked – and smelled – like a testosterone-production factory. He was big and well muscled, not at all lanky or weedy like some 'teenage' dogs. He was clearly every inch a male, with a huge, wide head, big broad chest and rock-solid thighs. He also had a very strong and distinctive odour – not something I'm inclined to tell owners as it tends to cause offence. It's not caused by the dog being dirty and needing a bath, though; it's caused by his raging hormones.

Even in the confines of my office, where the scent of other dogs is no doubt prevalent to the sensitive nose of the average canine, Toddy appeared to be 'high' on the hormonal effects of testosterone. He didn't even glance at me or his owner, preferring instead to suck up the smells around him. After he found one particularly delicious scent he seemed oblivious to being moved by the collar, and instead seemed to go into a kind of trance, chattering his teeth and drooling at the same time. This is called the 'Flehmen response'. The dog will draw a scent into his nose and mouth and over a special organ located in the roof of the mouth, called the Jacobson's organ. Bearing in mind that humans don't have this ability, all we can do is make a best guess that the dog is effectively tasting and smelling the odour simultaneously. This is nearly always a sexual response, and one commonly seen in male dogs that are 'tasting' the urine of a bitch to determine whether she is in season.

We took Toddy outdoors to attempt to break him out of his 'fantasy'.

Outside, Elaine explained that even walking Toddy on the lead was difficult. Despite numerous training classes and the fact that she felt sure he 'knew' how to walk to heel, Toddy would drag her from place to place, primarily so that he could cock his leg as high and as frequently as possible all over his chosen vicinity.

When we came back in and sat down, Elaine sighed deeply and said in a resigned voice, 'I know that you are going to tell me that I need to be the pack leader but I've put in so much work and I just don't know

what else to do.' She looked as though she might be about to cry.

I wasn't about to do that – which would at least come as something of a nice surprise. I asked Elaine whether she was planning on breeding from Toddy and what her feelings were about castration. This is always an important question. Whether or not I feel that neutering a dog will help in the case of a training or behaviour problem is utterly irrelevant if the owner has strong beliefs one way or another on this topic.

Elaine expressed surprise. 'But he's not showing any sexual behaviours,' she said. 'He doesn't mount other dogs or visitors or anything.'

It is a common misconception that mounting is the be-all of canine sex! This is often seen as the peak of the doggie experience, not the build-up to it. Mounting can also be an expression of frustration, the ultimate in attention-seeking behaviour and even a precursor to aggression – with no connection to sex whatsoever.

On the other hand, hormonally driven behaviours can include some which look like stubbornness, challenge and obsession. They can also have a more subtle impact, which is rarely considered, on the behaviour of *other* male dogs. High levels of testosterone, as commonly experienced by adolescents, can act as a red flag to other dogs. They may mark them out as the equivalent of football hooligans who are begging to be jumped on and beaten up by older males. Of course, this can easily lead to the poor adolescent becoming defensive or even aggressive in the presence of other dogs, if he anticipates that he's going to be set on for no good reason.

While an adolescent's desire to play with other dogs is completely normal, an obsession with play fighting can be 'charged' by testosterone. After all, it is the

hormone of competition and this – just to make matters more interesting – can become addictive, leading to its own set of problems, such as 'selective deafness' in the park and escapology from the garden as the dog is driven to seek a 'fix' of testosterone-fuelled fun.

Many studies have been done on the behavioural effects of testosterone – some of the more interesting ones on our own species. They've found that testosterone levels can be influenced by events, particularly what is described as 'success' and 'failure'. This was neatly demonstrated by some famous studies in which football fans had their testosterone levels measured before and after watching their teams battle it out in a World Cup final. The researchers found that the average testosterone levels of the winning fans increased by 27.6 per cent, compared to a 26.7 per cent *decrease* in those who supported the losing side.

Increased testosterone seems to be a widespread side effect of 'winning', whether this is vicariously via your football team or more directly through making a mint on the stock exchange. Such an increase is, in itself, likely to feel good – and this drives the person, or dog, to do it again. Some adolescent male dogs who like to spar with other dogs by wrestling and play fighting may well be repeating the behaviour because of this emotional 'high', which is difficult to compete with by offering lesser rewards, such as food treats or play with toys, that would normally work well in other circumstances.

All Toddy's behaviours added up to the same result. His drive to play with other dogs, his obsession with their scent messages and his need to leave his own by urine marking on every possible vertical surface outside all pointed to the fact that a reduction in testosterone would help his harried owner regain control over her boisterous boy.

I am not an all-out fan of castration for all dogs. As with every other aspect of behavioural intervention, it needs to be looked at on a case-by-case basis. There are lots of myths surrounding neutering and, astonishingly, many of the reasons given both for and against are without any solid scientific backing. However, in the case of Toddy I felt that neutering would not only help the owner but would do the dog a favour too. While testosterone-charged behaviour may be just a result of 'teenage' hormones and may settle down once the dog is fully mature at the age of about three years, until then he is pretty much at their mercy. If, during that time, the dog is frequently subjected to being 'disciplined' by other adult male dogs, he can soon become wary and defensive when meeting them – which in turn can lead to an aggression problem that could easily have been averted. Quite apart from this, neutering also prevents these dogs from experiencing the almost continuous frustration that unrequited hormones can cause. Frankly one can only imagine what this must be like.

In years gone by, the options available to the owner of a male dog in need of testosterone reduction were limited. Castration is generally favoured as it is permanent and surgically quick and simple, and it clearly prevents unwanted litters from being fathered. Chemical alternatives were generally regarded as unreliable, especially as they could have an unwanted sedative effect. These days we have another alternative – an implant called Suprelorin, which prevents the production of testosterone in males for up to six months and has been used for some time to prevent breeding in other species. Its effects are completely reversed after this time, making it a useful option for owners who are worried about the impact that castration might have on their dog's behaviour.

Armed with this information, Toddy's owner discussed the options with her vet and decided that

Toddy should be neutered. Her follow-up call six weeks later yielded some pretty surprising results. Not only was Toddy more responsive, less likely to head off in search of playmates and showing a dramatically reduced urge to scent mark, but all the training she had previously done with him had suddenly come back to the fore. Elaine was so pleased with his progress that she had even started doing agility training with him – something that would have been impossible while he still had his selective hearing problem.

Interestingly, I am more often consulted regarding the pros and cons of neutering male dogs than females. This seems ironic, considering that with bitches the surgery is more complex and the post-operative period more prolonged. Perhaps the thought of neutering a bitch is somehow influenced by the bits affected being 'hidden' – while castrating a male is all too apparent. While the effects of male hormones can be complicated by temperament, environment and external events, female hormones can be the source of some very complex social conundrums.

Photos of Trixie and Tilly arrived on my desk before I had even spoken to their owner. These beautiful

Bernese Mountain Dogs had lived together for nearly a year, playing happily and enjoying spending time together. Trixie was four years old, with a glossy coat and lolling tongue. At just nine months, and although a long way from being fully mature, Tilly was almost the same height as her housemate. The two bitches had come from the same breeder but they were not related, and everything had been hunky dory until a month previously when Trixie started growling at Tilly whenever she came into the room. While this had bothered Norman and Sheila, the dogs' owners, they hadn't thought much of it – until the day of the big fight.

'It was so out of the blue,' Sheila said, as I sat in their lounge a week later. 'I couldn't believe it. One minute I'm petting Trixie, and the next she and Tilly are having the most terrible fight. There was fur flying everywhere and we couldn't get hold of them. I thought they were never going to stop.'

Fights like this can clearly shake people up, and the bigger the dogs involved, the more frightening the scenario can be.

'Norman ended up running in from the garden,' she continued. 'He tried to split them up. We think one of the dogs accidentally bit him when they were going for each other.' Norman was nursing a bandaged hand.

'Apart from the growling, did you notice any other tension between the two dogs?' I asked.

'No, not really,' Sheila replied. 'They seemed fine together, although thinking back I suppose they haven't been playing much lately.'

The two dogs lay in the lounge with us, apparently unconcerned by the topic of conversation going on around them. Trixie lay by Shelia's feet, while Tilly lay over by the door. Both dogs might have appeared at first glance to be relaxed, but their body positions and the fact that they both seemed to be keeping one eye on proceedings told me otherwise. Norman shifted his weight slightly in the armchair opposite, and Trixie raised her head. Tilly lifted her head too, and watched, mouth closed, forehead furrowed. Just for a second, she went still. Trixie, having decided that no one was going anywhere, put her head back down. Tilly panted a couple of times, then put her head back down too, averting her gaze as she did so.

This was classic 'silent swearing': two dogs in the same house who are effectively giving each other the cold shoulder, and waiting for the other to make the first move. The tension was almost palpable between them and I could imagine the furore if they got going again.

While a semblance of peace remained, I thought I'd ask about the dogs' daily rituals, routines and diet, as well as having a chat about the hormonal status of the bitches. As a hobby, the couple enjoyed taking the dogs to local shows and proudly showed me the collection of rosettes they had won for their good looks. This meant that they hadn't had either of the girls neutered – it's conventional not to show neutered dogs – and that they were distinctly reluctant to do so. Tilly had not had a season yet, but I would have bet good money that she was about to come into season and that the resulting hormonal influence had caused the onset of the conflict between the two bitches.

Unfortunately, the phrase 'the female is more deadly than the male' goes for all kinds of species – and dogs are no exception. While aggression between male dogs in the same household nearly always seems to be focused around resources, such as food, toys, or access to owners, female conflict may start this way but soon becomes much uglier when the effects of hormones – in this case the equivalent of PMT – come into play.

While this might appear to be a dominance issue, there are other, more directly practical reasons why nature wants to create distance between dogs, and one of these is simple – space. Space is a fundamental resource for most dogs because it means safety. Humans know this instinctively. When someone stands too close to us we have an irresistible urge to move away. Places where we are forced to stand too close to one another – such as lifts or the underground – are distinctly uncomfortable, and it's no surprise that when you put too many people in a small space together tensions will often rise. Just think about the arguments many families have at Christmas, when they get along – albeit at a distance – the rest of the year. Bitches with puppies need space to feel at ease, and having a season is the first step to the bitch preparing to have a family.

Just how many dogs are too many for a space is such an individual matter that there are simply no rules. I have seen households where one dog that lived happily on its own couldn't cope with the arrival of another, even though there appeared to be more than enough resources to go around. I've also seen forty dogs that were rescued from living together in one caravan and, remarkably, none of them had any serious injuries from fights while they lived in such close proximity. Quite how they coped is anyone's guess, but the possibility of them all standing very still and avoiding eye contact, like too many people in a lift, springs to mind.

In Tilly and Trixie's case, the most urgent intervention was to ensure that Tilly's ever-escalating hormone levels did not reach a peak when she actually came into season. If this were to happen, the risk of permanently damaging the bitches' relationship would greatly increase. Hostility experienced for prolonged periods can simply become a learned habit. While some might have advised putting one of the dogs into kennels or having her looked after by relatives during the season, this would have run the risk that one bitch effectively believed she had 'got rid' of the other one – and the conflicts would only increase when they experienced the unwelcome return of the one that went on 'holiday'.

After much discussion, Tilly's owners consulted their vet and she was given an injection to suppress her season. Norman and Sheila managed the two bitches carefully for the next few weeks by making sure that they didn't inadvertently trigger off conflict between the dogs, keeping them calm and defusing any tension between them. They then decided to have both bitches spayed to ensure that there would be no repeat of their previous animosity.

Of course, neutering is not always the answer. It is not the panacea for all behavioural problems. If the issue is not related to hormones, neutering is likely to have little or no impact and, rarely, it can make behaviour problems worse. As we have seen, testosterone is the hormone of confidence and success. This means that the castration of dogs that are already lacking confidence, or those that are nervous, anxious or fearful, may in fact make them worse. Equally, there is some evidence to suggest that rather 'masculine' bitches – those that lift their legs to urinate, scratch up the ground after they have defecated, or who are over-bearing or aggressive with other dogs – may become even more androgenised after being neutered. This is primarily because the loss of female hormones means that the remaining testosterone – which bitches also produce – becomes predominant.

My belief is that many of the reasons owners are given for having their dogs neutered rather miss the point. There is remarkably little evidence to support claims that it has huge health benefits, and although it clearly prevents unwanted puppies from being born, this argument is more often directed at owners of non-pedigree breeds. Owning a pure-bred dog is not a justification for breeding from it. In my view, the only two factors for breeding should be exemplary temperament and physical soundness – not looks or an impressive pedigree. On the plus side, neutering can reduce some hormonally based drives, particularly those connected to competition, and it clearly reduces frustration in unrequited males.

As in so many other areas of canine behaviour, each dog needs to be considered individually rather than lumped together under one 'rule'. Of course, this takes time and consideration but I for one think they are worth it.

TOP TIPS ON UNDERSTANDING SEX AND HORMONES IN YOUR DOG

» Be aware that hormones can seriously influence behaviour. Think about the long-term consequences of keeping your dog entire or of neutering, not just the short-term ones.

- » Act quickly and take advice if you suspect that two or more of your dogs are in conflict. This may not be overt; tension and stress in the household can be the precursor to aggression.

- » There are many urban myths surrounding hormones and breeding. Bitches do not need to have a litter of puppies, and it is not advisable to allow male dogs to mate once or twice, as this just gives them a taste of what they are missing.

- » Many dogs put on weight after neutering, but this is not inevitable if you give them sufficient exercise and control food intake.

- » The timing of neutering can be all-important. Adolescent males may be picked on by other dogs if they smell like a testosterone-charged thug, which can lead to learned defensive habits. However, neutering too soon may have an impact on a dog's overall confidence, especially if he is already a bit of a wimp.

- » Neutering is not a panacea for all behavioural problems in dogs. Many problems to do with control or learned behaviours may be completely unaffected by the procedure. An implant that prevents the production of testosterone for up to six months can help to show whether or not the effects of castration in males will be beneficial.

» There are some cases in which neutering is not advisable. Very fearful or anxious young dogs may become worse after castration and dogs with a tendency to guard food may become more tenacious about this, not less. Seek behavioural advice before taking the leap.

Biology and behaviour; nature and nurture. Which has the greater impact, which is the most important? Are our dogs a product of their genetics, their hormones and their neurochemistry? Or is it their early rearing, training and environment that have most effect? Of course no one answer can be correct, simply because the question itself is wrong. Water is made up of hydrogen and oxygen. On their own, these two elements are just that – hydrogen and oxygen. It is only their *interaction* together that creates H_2O. This is a useful analogy when thinking about behaviour, and especially when thinking about something as integral to the dog as hormones.

Testosterone in particular is often regarded as the 'cause' of aggression. Sure, there is definitely a high correlation between testosterone and aggression, but it's important to realise that this is a correlation, not a cause. Study after study on all kinds of animals – including humans – has shown that aggression increases testosterone secretion but not that testosterone necessarily increases aggression. Aggression has to be a pre-existing state for testosterone to exacerbate it. This is interesting to me as a behaviour specialist, and to the thousands of owners of dogs with aggression problems, because it raises questions about the efficacy of neutering as a 'cure'.

Castration in male dogs does, of course, dramatically reduce testosterone in the system and it prevents further secretion but it does not alter previously learned patterns of behaviour, or a genetic pre-disposition to use, or enjoy, aggression. You may conclude that castration in male dogs is less useful than your veterinary surgeon might suggest, but if you hesitate and delay the procedure too long, so that the dog has had a chance to practise aggression prior to castration, the behaviour can become ingrained or reinforced.

Like so many areas of behaviour work, to me this issue is about treating the individual as an individual. Castrate a male dog pre-adolescence and you might be left with some pre-pubescent physiology, such as under-developed muscles, or puppyish behaviour, such as appeasement urination on greeting. Wait until he has learned to use aggression as a means of bullying other dogs or getting what he wants, and you may wish you'd acted earlier – the behaviour is learned. Along with knowing your dog, clearly timing is everything.

Part 4

Happiness

JVK

10

Life with purpose

Going with the flow

Case history: **Bella, the flame-herding Border Collie**

It's perhaps ironic after so many years of humans benefitting from animal studies that the reverse is now also true. Some of it has now come full circle and recent advances in human psychology have allowed us to map across many of the basic patterns of emotionality to dogs without being accused of 'humanising' our pets. Of course, the fact that dogs experience a whole range of emotions comes as no surprise to the average dog owner. Clearly, dogs show anger, sadness, joy, attachment and even depression – all emotional states that we can recognise and empathise with. However, as dogs are unable to control their environments or their circumstances, their emotional welfare is often as much a human responsibility as giving them food or water.

While all pet dogs should ideally have access to all kinds of things that give them joy and fulfilment, a broad base of contentment should come first. Basic contentment depends on emotional attachment, an outlet for natural 'hard-wired' behaviours and the opportunity to learn how to fit into a human-driven existence without conflict. Where these are not present, problems are bound to follow.

I first saw Bella on a clip of home video footage sent to my mobile phone. If it hadn't come straight from the client, I don't think I would have believed what I saw. There was a lovely, if rather skinny, Border Collie bitch leaping straight into the flames of the family's real-flame gas fire. Even in the slightly grainy footage I could tell that she was making contact with the glowing 'coals' and was definitely putting her paws right into the flames as she pounced in and then leapt back.

I called the family immediately. Their description of her behaviour was almost as dramatic as the actual pictures. Bella was eleven months old, barely beyond puppy stage. Right from the outset, she had been a 'difficult' dog. Bought as a replacement for an elderly family dog that had died the previous year, Bella was clearly a bit of a shock to the system. I asked the owners to put together some history for me and get a referral from their vet, and I arranged to make a house call.

Bella greeted me at the front door with enthusiastic wagging and much happy leaping and wriggling. She tried to grab my hand as I went into the living room, and the owners immediately shouted at her and attempted to pull her away. They were clearly embarrassed by her behaviour but I reassured them that I'm used to all kinds of welcomes – many of them not nearly as friendly as this one.

Once seated, I went through my usual routine of ignoring the dog, and asked her owners to do the same. This has nothing to do with dominance but everything to do with safety. Until I have observed and assessed a dog fully, I never expect it to be friendly to me. Why should I? Trying to touch or make friends with an unsocial or fearful dog can be a fast route to the local hospital, and it's poor canine manners to touch a dog unless the dog invites you to do so. The other reason why I tend to ignore dogs at the start of a consultation is that it gives me a chance to see what the dog does once the owners stop trying to make a good impression on me. When the owners are calm and settled, does the dog calm down or does it become more anxious or excitable? Does the dog choose to be with his owners or am I more of a draw? I watch the dog like a hawk during this time. Even where the dog sits or stands gives me valuable information about their relationship.

I took my pen and notebook from my bag, and started chatting with the family. John and Jean were in their sixties. They had kept Border Collies all their married life and were keen to tell me just how unlike their other dogs this one was. During the day, Bella was described as hyperactive. She was on the go from the minute she heard the couple get up in the morning until the minute they put the lights out at night. Her antics would include stealing food from the kitchen surfaces, barking at people and dogs going past the front window, throwing herself at the front door when the postman arrived, pacing round the front room and leaping on the furniture. When out for a walk, Bella's behaviour wasn't much better. She strained on the lead so badly that she had pulled Jean over on a number of occasions, meaning that now only John could take her out. The couple had only attempted to let her off the lead a couple of times because when they did she had seemed terribly distracted and they had trouble getting her to come back when they called.

During my conversation with John and Jean, I watched Bella carefully out of the corner of my eye. This was one seriously bright dog. After failing to get my attention by climbing on my lap, she made a couple of half-hearted attempts to mouth my hands. Finding that this didn't get the usual reaction, she carefully put her teeth round the end of my pen and tried to take it from me. Thwarted, she inserted her velvety Collie nose into my bag, pushed the zip open by working her muzzle inside and pulled out a tissue. Delighted with herself, she headed off to her bed in the corner to demolish it. Her owners were almost beside themselves with consternation. Their dog was behaving like a delinquent and we were doing nothing about it. They practically had to sit on their hands not to get up and remove the tissue, reprimand her for her behaviour and take some steps at least to feel as though they had a semblance of control. But I had other ideas.

Bella ate the whole tissue – an unremarkable and harmless incident in itself – then returned to the scene of the crime. She had found a trophy there once before, so she came straight back to my bag and was surprised to discover that the bag was now zipped completely shut. She examined it closely, looked at it from all angles, pawed at it and gave it a small chew. Over the years, I have tested all kinds of bags for dog-proofness. This was one of the best. Made of strong stuff, it wasn't about to give in to the attentions of an adolescent Collie – and neither was I. Once she realised that no more fun was to be had from the contents of my bag, Bella stood stock still in the middle of the room, looked at me, then at her owners, and took a huge spring straight into the flickering flames of the fire! In she leapt, then back, then in again, biting at the flames and retreating – all in the blink of an eye.

Involuntarily, we gasped. Bella leapt back and raced around the living room in an athletic circuit. Her owners jumped up and ran after her, attempting to catch her as she dodged around them using the lounge furniture as an agility course. Bella was fast. Ears up, tail flying and tongue hanging out, this wasn't a dog suffering from anxiety, but rather one that was – in the moment at least – experiencing something akin to an ecstasy high. In and out of the furniture she wove, clearly playing a mad game of tag that centred around her attempts to leap back into the fire without being caught by her owners.

After some time (only minutes, but it felt like much longer) they captured the now-panting and over-excited Collie and restrained her on a lead, whereupon she flopped to the floor with a deep sigh of resignation.

'You see?' said John, red in the face from exertion. 'She's mad.'

Jean nodded. 'We can't cope with much more of this,' she said. 'If she can't be cured, she'll have to go.'

I looked at the dog. Bella was lying on the floor in classic Collie pose: mouth open, panting and tongue lolling. She clearly didn't think anything was wrong; indeed, she'd just had the most fun a Collie could have in her whole day.

While her owners may have thought her behaviour was unhinged, to her it was purely and simply a motor pattern: no more and no less than any other motor pattern in the animal world. Motor patterns are behaviours that are hard-wired, inherent and instinctive. They are the patterns that make us act in the way we are born to. They are the patterns that make cats pounce on mice, that prompt birds to peck or worms to come to the surface of the soil when it rains. Bella's behaviour wasn't mad; it was perfectly in keeping with her breed and natural drives. Bella wasn't just jumping into the flames – she was herding them.

There can be little doubt that herding dogs – and Collies in particular – often have a hard time coping with a pet home simply because their drive to do what they were bred for is stronger and closer to the surface than in other breeds or types. It's a sad fact that in my behaviour practice many of the dogs we see for behavioural problems are simply suffering from being in the wrong environment. Put a dog with a strong herding instinct into a home with small chil-

dren, for example, and instead of herding sheep, the dog may well try to herd the kids; then, of course, we call it chasing or aggression, not natural instinct.

So it was with Bella. With no occupation for her problem-solving brain and no outlets for her natural drive to herd, she had discovered that herding flames was a pretty good substitute – and it got sensational attention too.

The fact is that dogs that have been bred for many centuries to herd, retrieve or bark don't simply give up these drives just because they are in a pet home. Why not? Well, for two main reasons. The first is that the hard-wiring in the brain which governs the behaviour cannot simply be shut off. It is just as fixed and permanent as coat colour or ear length.

The second reason why dogs do not just give up behaviours they have inherited from their domesticated bred-for-work ancestors is because the effects are quite literally 'in their blood'. You know this from watching a Spaniel quarter a field, zig-zagging across the ground, looking for a fallen bird, or observing a Greyhound running – they love it. When dogs are engaging in a natural motor pattern, pleasure chemistry courses through their bloodstream, body and brain and, it could be said, quite literally fills them with joy. Of course, humans know and recognise this experience: the feeling of being completely caught up in some skill or task that you are good at, and that you enjoy. The 'work' is effortless, the challenge exciting, and all sense of time is abandoned. Human psychologists call this 'flow' and they recognise it as an essential part of human happiness. In my view, it's about time we recognised it in dogs too.

Bella was clearly a dog in need of more appropriate outlets for her hard-wired drives. The pleasure that she was experiencing in herding flames was clearly going to get her into serious trouble. I asked her owners what they had done already to stop this behaviour, and they said they had tried almost everything. Of course, any attempts to punish this behaviour would be completely ineffective; motor patterns cannot be switched off – only redirected. Her owners had even tried living without the fire on. It was cold but they had persisted for several days, only to discover that Bella would jump into the grate as if the flames were there anyway, forcing them to take action.

Although a mystery to Bella's owners, this is perhaps not surprising. Just like humans, dogs strive to create the feelings of pleasure that they so enjoy. In fact, in the absence of something to herd, Bella only had to *imagine* the moving, flickering flames in order to trigger the motor pattern response that released the pleasure chemicals in her brain. Maybe this is just the equivalent of doggie fantasy?

Bella's needs were relatively simple: more exercise, more mental stimulation and a chance to express her natural drives but to put them into practice was another matter. As with many of my cases, we had to start small and take one step at a time. Over a series of weeks, we taught Bella to walk nicely on the lead. This at least meant that she could get out more, and accompany her owners on trips into the local village to collect the morning paper. How to retrieve a ball came next. This helped with her training off-lead, and meant that she could run free in the safety of the local park. Chasing a toy is the number one need for most herding dogs that don't have access to their own sheep. It allows them to engage in all the natural behaviours they use when they are herding prey animals – stalking, chasing and grabbing – and it gives them the same pleasure burst too.

Bella's considerable Collie energies were also channelled in a more constructive fashion by an unexpected source when John and Jean's fourteen-year-old niece asked if she could take Bella to agility classes.

With a bit of adult help, she and her canine team-mate had soon mastered the basic obstacle course and were flying round the field with ease.

Of course, as we all know, pleasure is addictive, and Bella's behaviour was clearly not going to disappear in a fifteen-minute-fix. At home, despite the fire staying off temporarily, Bella continued to attempt to leap into the 'pretend' flames of the extinguished fire for many months, much to her owners' exasperation. However, over a period of time, they learned to ignore the behaviour, which prevented her from being rewarded by their attention, and very gradually, as her other outlets for physical exercise and mental stimulation prevailed, Bella's obsession with the imaginary flames seemed to diminish. Once this had subsided, we could instigate a new fireside routine for Bella – in the shape of things to chew.

Working on the basis that Collies love patterns, habits and predictable sequences, we gave Bella a small chew each and every time her owners were going to sit by the cold fire. Once this pattern was established, her owners would make as if to click on the ignition for the gas fire, before immediately giving her the chew and settling down to read the paper or watch TV. Being a smart dog, Bella worked this out within a week and would sit and look at the tin where the chews were kept rather than paying attention to the fire when her owners clicked the switch. After this came the real test. We walked Bella, tired her out by playing with the ball, and brought her back home, where she had her dinner. Next, we headed into the living room, clicked on the fire and turned it to a low heat, sat down and gave Bella her chew, whereupon she settled down happily to demolish it, without even a glance towards the flames.

Bella is by no means an extraordinary case. Although the outlet she chose for her hard-wired motor patterns was extreme, there are many hundreds if not thousands of dogs all desperately attempting to fulfil their working potential, often in the most bizarre and unusual ways.

How many Terrier owners give their dogs a place to dig? What happens when the modern Border Collie tries to herd up joggers or cyclists? How can the average Golden Retriever cope with living in a home where picking up the kids' toys is frowned upon? If you already own a breed or type with a high drive to perform a certain behaviour, it is up to you to provide an outlet for it. If you are considering getting a dog (or another one), then think carefully about your circumstances and environment, and whether they fit with your chosen breed's characteristics. This may not be as obvious as it sounds: the 'fit' does not depend on size, but on behavioural needs. For example, all the working breeds are doomed to a lifetime of pent-up frustration if kept behind closed doors, fenced in or tethered. They need space and exercise and a huge amount of socialisation and training to ensure that they are safe around people and other dogs. This goes just as much for Jack Russells as it does for German Shepherds.

Hard-wired behaviours and 'motor patterns' cannot be 'trained out', eradicated or suppressed. These behaviours are a basic need for many breeds, prompting pleasure chemistry that borders on ecstasy. Our challenge is learning how to give a safe and practical outlet to these natural, if inappropriate, behaviours in our modern society. After all, don't our dogs deserve to be happy too?

TOP TIPS ON MAKING SURE THAT YOUR DOG HAS PURPOSE

» Know what your dog was bred for. There's no excuse for buying a specific breed and then being surprised when it acts the way it was bred to.

» Give your dog lots of appropriate outlets for its natural behaviour and instincts. It can be a challenge to think of ways to do this, but it's essential.

» In order to control your dog's natural impulses, you will need to invest time and energy in training and exercise. Don't underestimate this.

» Many of the working breeds are highly intelligent. Twenty minutes of 'mind training' can be the equivalent of an hour of physical exercise – so lots of mental stimulation and training are required.

» Agility training and flyball – where dogs relay race over hurdles to catch a ball and then return to their handlers – are just two examples of dog sports that require discipline, self-control and responsiveness from the dog – and also offer an excellent outlet for pent-up energy.

» Some environmental influences – such as diet, a noisy or chaotic household, or being cooped up in a small area – can serve to increase a dog's frustration at not being able to express natural 'hard-wired' behaviours. Try to see all this from your dog's point of view.

Trap me in a hotel room with only a TV for company, and I would probably last one day before becoming irritable. Any more than two days and I could be potentially dangerous! We understand how frustrating boredom and inactivity can be for humans, and yet over and over again I meet high-drive dogs that are meant to live happy and contented lives in the mental equivalent of that hotel room. This is not confined to the typical scenario of a Working Collie living in a tiny upstairs flat while their owner goes out to work all day. Rather surprisingly, the dogs I see that are most likely to suffer from lack of stimulation are those whose owners think one walk a day and access to their large garden will be enough. Dogs – especially those with a strong work ethic – will start to demonstrate all kinds of self-reinforcing behaviours in response to being kept without enough to challenge body and mind. This can easily lead to detrimental emotional states too. No wonder such dogs have a short fuse. No wonder it then takes very little to trigger an over-reactive response to a relatively trivial trigger. And then, just to add to their woes, they are labelled as dominant!

Having things to do is a fundamental part of living a fulfilled life. Acting out motor patterns in a socially appropriate way is an essential part of contentment; the fact that humans have invented crossword puzzles, dancing and golf is surely testament to this. Dogs need to express natural behaviours, and most especially those that are hard-wired motor patterns. Trying to prevent these from being displayed is practically impossible; suppressing them in one area means that they are very likely to pop up in another.

While exercise is a fundamental part of keeping a dog healthy and happy, simply toning the body is not enough. Indeed, for some bored dogs, increasing hard exercise through running next to a bicycle, or on a treadmill, simply means that their stamina and fitness improves – and they have more energy to put into relieving their boredom. Dogs need to sniff, they need to have social contact with people and other dogs, they need to solve puzzles. Dogs need to have their minds challenged and to enjoy being part of a team with their social group. Dogs need time and input from us – and surely that's why we have them in the first place, isn't it?

11

Protection and connection

Park life

Case history: **Marie and Morris**

I'm watching one of my dogs lying on the carpet. She has had a busy day and is flat out and fast asleep: paws twitching, muzzle crinkling and apparently enjoying a pretty good dream in which she seems to be chasing squirrels. Of course, she could be re-enacting some kind of canine superhero fantasy, but who can tell? To me, this is both the eternal dilemma and the endless pleasure of keeping dogs, because when she wakes up for dinner in an hour or so, she won't be able to tell me, as a person would. The inner workings of dogs' minds remain a mystery.

This means that we have to make best guesses about dogs' like and dislikes, thoughts and feelings, based on their observable behaviours, body language and facial expressions. The problem is that we have to try not to filter these observations purely through our own experience – and this can be tricky when the human and the canine worlds collide.

Ask most humans what they think makes their dog happy and they will give sincere, if rather boring, replies: 'Well, he likes eating his dinner', 'Being scratched behind the ears', 'Going for walks'.

Yes, but ask a dog what presses his jolly switch and I bet you'd get a whole raft of different responses – and not all of them high on the list of behaviours that humans find appetising or even appropriate. For example, one of my dogs considers rolling in fox poo to be the equivalent of winning the doggie lottery.

Of course, dogs need exercise, and because of this it's tempting to think that by putting the lead on and taking your dog to the local park for half an hour you are providing what your dog wants and needs. However, if you asked many dogs what they think of this daily routine, you might be surprised to hear the answer.

Marie was one of the most devoted dog owners you could wish to find. Besotted with Morris, her three-year-old black Labrador, she made sure she gave him everything she thought he could want: a secure home, training and plenty of attention. Aware of the fact that they lived in London and that she only had a small garden, she was diligent about giving Morris the right amount of exercise, and made sure she got him out for frequent on-lead and off-lead walks no matter what the weather or her own routine. It was therefore an increasing irony that the very walks in the park that should have been giving both dog and owner daily pleasure were in fact becoming a nightmare.

At the point when I met Morris and Marie, the situation had become so stressful that she just didn't know what to do. Daily off-lead walks were becoming a hot-bed of conflict, with the normally gentle and lovely Morris getting into fights with other dogs – always a terrifying experience. Despite the fact that there had been no serious injuries following these confrontations, Marie would frequently return home in tears, only to dread the same scenario happening all over again the following day.

Sadly, this is not an uncommon situation. On the whole, we are pretty lucky to be able to walk our dogs off lead in woodland and parkland all over the UK, but this does mean that we are constantly relying on the good nature and training of the other dogs we meet and of the responsibility of other owners. In my experience, it only takes a few unpleasant encounters to trigger a pattern of behaviour in both dogs and owners that is counterproductive to the whole 'social scene' in the area. The way your dog sees the park may not be quite how you see it – and once again, thinking from your dog's point of view is an important part of being a good owner.

Parks and dog-walking areas vary enormously. In one glorious place where I walk my dogs, the only other living creatures I'm likely to see are deer. Beautiful for me and, of course, absolutely fantastic for the dogs – except that they are not allowed to chase them. I have put in two years' worth of training to prevent my dogs doing this, using kind, non-confrontational methods – and certainly no electric shock collars, which many other 'experts' would reach for in the circumstances. In another area where I go regularly, I meet literally dozens of other dogs. This is great for socialisation, but only with dogs that are truly social. I regard it as my duty to protect my dogs from those that aren't, and this involves being able to spot which ones are to be avoided as well as which ones will be good company.

In my role as a behaviour specialist, I don't just sit in an office and dole out good advice; I'm a hands-on trainer and proud of it. This means that in most cases where walks are an issue, I accompany the owner and dog on an 'average' walk to the park or woods, or wherever they usually go. Frequently, I am horrified at what well-meaning and loving owners will subject their poor dogs to in the name of freedom and exercise. Time and again, dogs are placed in situations they can't handle, or are forced into confrontations with other dogs that, given a choice, they would avoid like the plague. Time and again, the owners I'm with are utterly surprised by what I tell them – and amazed when I recommend that they supervise and intervene on walks in order to help their dogs and prevent conflict.

For some reason, our doggie 'culture' dictates that once dogs are off the lead, they should be left to their own devices. Such idealistic attitudes from owners are not confined to the UK. In many other parts of the world where dogs are kept as pets, the issue of freedom and exercise is also a contentious one. In many areas of the USA, dog parks are commonplace in towns and cities and a quick peek into pretty much any of them is enough to make my hair curl. In these fenced and restricted areas too many dogs have to face too many head-on confrontations each and every day. Watch carefully and you'll start to see what's really going on. There are dogs standing on the park benches. Is this because they are trying to be dominant over the ones that are standing beneath them on the ground? No, it's because they are frightened and desperately trying to stay out of the way. Ah, there are two dogs having a lovely play – one is chasing the other round and round. Their owners chat away on mobile phones, completely oblivious to the fact that one dog is aroused and predatory while the other is running for its life. Dog parks don't come high on my list of places to take dogs – indeed, with the rare exception of those that are well run and managed by people who truly understand dogs, they aren't on my list at all.

'Yes, but ...' I hear you cry, 'we don't have dog parks in the UK.'

Perhaps not yet, and perhaps not in the form that they exist elsewhere, but I see many of the same kinds of problems in urban parks all across our fair land. Marie and Morris took me to one. Marie said it was worth the car ride because it meant I could see how Morris was in the 'big park'. We arrived after a twenty-minute traffic nightmare. I looked for the big park but all I could see was wire-mesh surrounding some old tennis courts. As we got Morris out of the car, I could almost taste his adrenaline. Already, his body was on high-alert: tense and aroused, head high to catch the scent of other dogs, threat and danger. He dragged Marie to the fence line and cocked his leg as high as possible, urine marking and scratching up behind him in an attempt to reassure himself that he was a big warrior, macho and invincible, ready for the inevitable confrontation inside the fence. I wanted to go home right then, even before we'd stepped one foot into the place, but it was not to be.

We opened the gate. The areas that had once been tennis courts were laid to grass and some hard standing, but with only one or two trees at the perimeters. The whole area was an open landscape, with nowhere to run and nowhere to hide. Without a second thought, Marie shut the gate behind her and let Morris off the lead. There was a group of lads at the far side of the area. Two dogs – a Staffie-type and a big mastiff – were chasing each other in a tight circle, barking. Aroused play or, in other words, fight practice. Morris looked at them, then turned and worked his way along the inside of the fencing, sniffing and urine marking as he went. The Staffie-type broke away from the group of boys and ran, in a straight line, hard and fast towards us. Half the park length away, the dog stopped dead. Morris looked up and saw him. The dog's body was tense and upright, tail up, stiff and unmoving. Like something out of a Clint Eastwood movie, Morris mirrored the other dog's posture, ran towards the other dog and also stood still. They were only inches away from each other and the jury was out. My heart was in my mouth. The tension between them was palpable. Marie shouted and started towards Morris, her hand reaching out to try and grab his collar. I shouted and put out my hand to stop her, for it is in these seconds when two dogs are squaring up to one another that intervention can actually trigger a fight rather than stop it. I beckoned to her to come with me and quietly move away, rather than towards the dogs. Clearly this was against all of Marie's instincts, and we watched nervously as the two male dogs walked on hot coals towards one another, sniffed for what seemed like eternity and then broke away to urine mark and scratch the ground like competing rhinos. It had been a close thing.

Marie and I turned and Morris followed. The Staffie-cross turned too and trotted back to his owner, who was blissfully unaware of the whole incident. Just as I was about to explain that prevention was all-important in protecting Morris from such near misses, and that perhaps Marie should put him back on lead for a moment, the gate flew open and five dogs exploded into the park. Without warning, they hurtled towards us. Morris, clearly still pumped full of adrenaline from the previous encounter, jumped on the first dog that reached him and with what can only be described as a roar, pinned it to the ground.

Thankfully, the hapless if rather clumsy Spaniel that found itself under Morris in that split second was shaken but unharmed. We got Morris back on the lead, apologised profusely to the dog walker, who waved an unconcerned goodbye as her charges headed off towards the group of lads on the other side of the park, and decided to call it a day. Poor Marie was in a terrible state.

'Why did he do it?' she cried. 'I can understand him wanting to fight another dog if it's aggressive towards him, but why was he so horrible to those dogs who didn't do anything?'

Human thinking. Canine feeling. This is where the two worlds collide. Despite the fact that we are probably the most violent species on the planet, to us aggression is something that we have to rationalise, to justify. To dogs it is a measure that they will generally avoid, but will resort to if their emotional state dictates. Poor Morris didn't really want to fight, he didn't even want to cause trouble, but put any creature in a ring-fenced area in which they feel threatened, and they will start to gear up for self-defence. Just like us, dogs have limited options when backed into a physical or emotional corner and the real risk is not that they will lose the first few fights, but that they will win them.

In my experience as a behaviour specialist, the worst kinds of cases I am asked to deal with are those in which the dog has learned that using aggression to intimidate others is not only successful, but also enjoyable. These canine equivalents of the playground bully may start out being fearful of other dogs or people, but once they have used aggression to get rid of other dogs, they quickly discover the powerful and triumphant effects of 'winning'. Such effects are well documented in human science as acting at a neurochemical level. In Chapter 9, we talked about the supporters of winning football teams who have higher testosterone levels than the fans of the losing side – which makes them feel like conquering champions. Victors can become addicted to victory and bullies can get hooked on beating up dead-cert losers – in the case of dogs, those that are smaller, older, weaker or less confident than themselves.

Morris was by no means a gun-slinging cowboy bully – yet – but the need to move fast to change his attitudes towards other dogs was pressing. First and foremost, I recommended that he was neutered. This was primarily to reduce the likelihood that he would be picked on by all the other macho males in the neighbourhood, as well as to ensure that he wasn't enjoying a testosterone-fuelled high in situations where he might pick on other dogs less confident than himself. However, in addition to this, my number one drive was to re-educate him about what the presence of other dogs meant, and this was going to require the re-education of his owner too.

The fact of the matter is that confrontations between dogs are frightening for people. Even when no injuries occur, the speed and noise of a dog fight is enough to spike most people's adrenaline levels off the scale. Our instinct may be to intervene, but it's a fact that we nearly always make dog-to-dog conflicts worse by getting involved. After all, as a slow and lumbering human, you are basically a hindrance to a dog in a fight – even your own – and it's little wonder that people often get bitten trying to break up dog fights yet discover afterwards that neither dog had so much as a scratch on it. Because of this, prevention is much better than cure. Teaching owners how to 'predict and protect' their dogs when out walking with them is far easier and much safer than attempting to break up canine fisticuffs that are already in full swing.

Over the next weeks and months, Morris's owner followed a five-step programme.

1. Change the environment

First, we needed to give Morris exercise and freedom but without the continual risk of meeting unsocial dogs in a confined and stressful area, especially one that he associated with threat. Marie changed his exercise regime to include more relaxed on-lead walks, more play with his tennis ball and fewer but longer trips to a large open space where there were lots of trees, no fences and no gangs of kids with uncontrolled dogs.

2. Create positive associations with seeing other dogs

Next on the list was to rebuild Morris's associations with the presence of other dogs. Marie picked up clicker training like a pro and was soon able to click and treat Morris for being calm and relaxed when another dog came into view, initially when he was on the lead walking in the street. This took time, patience and generosity with the treats, but he turned out to be the model student. Marie also learned how to spot when Morris was becoming aroused – his ears went up and his tail stiffened – and simply to stop and wait until he calmed down again rather than exacerbating his tension by tightening the lead, or getting worried herself.

3. Meet nice dogs and be relaxed around them

Morris also needed to make and maintain some canine friendships. Thankfully, Marie met several lovely owners at the open space who had dogs that were not intimidated by Morris's rather impolite initial approaches. Adult bitches are generally a good choice for this as they don't suffer fools and yet are confident enough to ignore macho displays of insecurity. With Morris on a long lead to ease the initial introductions, he was encouraged to lower his head and body posture by foraging for treats which Marie had thrown into the grass for him to find – which instantly gave the other dog reassurance about his non-threatening intentions and got the meeting off to a good start.

4. Predict

Finally, and perhaps most importantly, I spent time with Marie teaching her how to predict other dogs' behaviour in order to protect Morris. This is a skill that anyone can learn and is, I believe, an essential component of working as a team with your dogs to keep them safe and happy. Walking your dog is not a passive affair. Indeed, it takes concentration and energy to stay one step ahead of the game. Don't even think about talking on your mobile phone, and only walk and talk with someone else if you can multi-task and stay alert to the environment. Ideally,

Here's what to watch for

Good signs	Caution!
The other owner spots you and your dog and is clearly attentive and making predictions of their own. They smile at your dog and look relaxed about the two dogs approaching or meeting.	The other owner is anxious about their dog's behaviour. They are ineffective at recalling their dog, or inattentive. They start shouting at their dog or shout at you!
The dog or dogs look relaxed and their movements are flowing and flexible.	The dog or dogs look stiff and tense.
The dog may spot yours but is happy to carry on sniffing the ground or continue with his or her previous activity.	The dog spots yours and immediately becomes fixed and focused on your dog.

you should be aware enough to spot another dog – or dogs – before your dog does. Once you have, you need to assess their behaviour and body language to ascertain whether or not you think they are safe and social enough to meet your dog. Obviously, your decision also needs to be based on a sound and honest appraisal of your own dog's behaviour and likely responses.

Good signs	Caution!
The dog is attentive to its owner.	The dog completely ignores its owner.
The dog comes towards your dog in a big, wide curve.	The dog heads directly for your dog in a straight line.
The dog slows down or stops before meeting your dog.	The dog charges at your dog.
The dog turns his or her head away from your dog when up close. He or she offers gentle sniffing at head and muzzle before moving on to sniff the tail end.	The dog keeps his head and spine in a straight line when meeting your dog. He puts his head or paws over your dog's neck in a classic 'T' shape.

5. Protect

There are a number of measures you can take to protect your dog from the advances of less than sociable dogs, and this is never so vital as with puppies and adolescents. The greatest of these is to spot potential trouble and avoid it. Your dog will thank you a million times over for simply taking a different path, retracing your steps to keep out of another dog's way or waiting off the track for a couple of minutes to allow a dog that doesn't look friendly to pass. Even if your dog is great with other dogs, you should assess the other dogs and make decisions based on your observations of them and their owner's behaviour. Trust your gut instinct and err on the side of caution. Once you have established safety, you can always use that most divine of human skills and talk to the other owner – and then allow the dogs to play. This is not the action of the play police; it is the behaviour of an owner who understands the need to protect their own dog and to ensure that positive associations are built and made with as many lovely, social hounds as possible. That's teamwork.

TOP TIPS FOR SAFE AND ENJOYABLE WALKIES

» Switch off that mobile phone! You can't concentrate on your dog if you're talking on a mobile, and your dog will know it.

» Learn to watch other dogs and predict their behaviour. Their owners will often give you clues as to their dog's likely reactions too. Don't walk in areas where there are so many other dogs that you can't assess them before they approach.

» Take action to protect your own dog and prevent conflict with unknown dogs. Walk around other dogs in a wide arc if you meet them while out and about. Try not to walk directly towards them or make your dog pass them on a narrow path.

» While it's usually best not to get embroiled in a canine conflict, I don't know many owners that wouldn't try to save their pet in an emergency. This is extremely risky, so always try to engage the other owner in helping too.

» If your dog is fearful of unknown dogs, teach him or her to stand behind you for protection. This is better than running away, which can trigger a chase response in the other dog.

» If another dog approaches your dog in a threatening way and it is too late to avoid it, walk away from your dog, not towards him. Don't reach for his collar or shout – you may trigger or escalate the conflict.

» Don't walk in the same park or field each day. Different walks – some on lead, some off – will give your dog interesting variety.

» Don't allow your dog to form patterns of conflict. It's akin to craziness to walk your dog past the same garden gate each and every day when you know the dog that lives there is going to come flying out, spitting and swearing. Take a different route to avoid known trouble.

» Do allow your dog to interact with and play with other safe, social dogs, but never to the exclusion of human interaction. Your dog should think you are the best playmate in the world – not another dog.

» Are you fun on walks? Would your dog think so? Play, give treats, train and interact with your dog during walks – this is not a time to be passive.

» Call your dog to you frequently, reward and then let him go again. This will prevent him having selective deafness when it's time to go home.

» You don't like every person you meet, so why should your dog love every other dog? Be realistic. If your dog takes a dislike to another dog, it's nothing to panic about. However, if your dog is being genuinely aggressive to other dogs, be responsible and honest about it. Keep him or her on the lead and seek help; it's not personal.

If ever there was a time when your dog needed to know that he is part of a team, it is when you are out walking. Away from the security of home soil, your dog needs your input to help him steer a path that is both safe and enjoyable.

Some dogs are excellent at reading other dogs' body language and facial expressions. It's likely that they are also fluent in reading olfactory signals – something that, as humans with lesser scenting abilities, we rarely think about. However, for many reasons, other dogs may be poor at noticing peer signalling. It may be that they came into the world that way – that the ability to read other dogs is partially genetic – or it may be that they had little opportunity to learn this skill when they were puppies. Of course, some dogs have had bad experiences with other dogs, and this can set up a life-long hatred of all other dogs – or just one type of dog. It makes sense, of course; being attacked by one large brown dog is likely to make the victim wary of other large brown dogs in the future.

If your dog seems socially unaware, or has had a bad experience in the past, the help you give him out on walks is crucial to prevent him becoming a victim once more and learning to use defensive behaviour to protect himself in future. Practise your assessment of other dogs' behaviour so that you can learn to spot canine trouble-makers from afar and take steps to avoid them, as well as weighing up potential new buddies.

If, on the other hand, your dog is possessed of excellent 'dog reading' skills, then it's important that as his team-mate, you go along with his superior decision-making abilities in this area when out on walks. This may mean that you need to follow your dog – for example, on a path where he chooses to give another dog a wide berth – rather than you deciding which route to take. Now, according to some dog books, allowing your dog to 'dictate' the route means that you are not acting as the 'leader' – but this is crazy. Would a 'leader' want to lead their 'pack' right into danger? Being part of a team means looking out for all the members in it, to keep them safe from harm. It also means understanding when sometimes your dog knows better than you.

JVK

12

Learning

What gets rewarded gets repeated

Case history: **Roxy, the ghost hunter**

Why do dogs do what they do? If dominance is not to blame (and I truly believe that this is a human construct which merely serves to make us feel better about our own behaviour) then only the power of genetically driven behaviour and learning is left.

Over the years, many studies have shown that the power of rewards and punishments makes slaves of us all. Of course, we like to think that we have free will, and so we do, but our past experiences and the impact these have had on our patterns of behaviour cannot be ignored. Some of the research is, frankly, enough to make you cringe, but it is interesting none-theless, and some of it may explain why despite our desire to have a good relationship with our dogs, we may still be tempted to use punishment if we are told to do so.

In 1961, the social psychologist Stanley Milgram embarked upon a series of experiments to examine the impact of authority on human behaviour. He engineered a study in which one person, the 'subject', was instructed to deliver increasingly severe electric shocks to another, the 'victim', if they made a mistake in a learning test. During the experiment, Milgram concealed the fact that the shocks were not actually being administered. Instead, the 'victim' would give an Oscar-winning performance as if they had been given a shock – enough to convince the subject that they really had delivered it. The results were as shocking as they were distasteful. Despite the fact that the subjects taking part in the experiment were horrified by their role in delivering apparently severe and painful shocks to the other person, about two-thirds continued until they were delivering the maximum level of shock, which was marked as 'dangerous' on the dial. Their justification? They had been told that the experiment must continue.

Although studies like this might make us lose confidence in our ability to act independently, it is the ability to learn – and learn fast – that is probably the most vital component of our relationship with dogs in the modern age. It is also something to bear in mind next time you are told by an 'authority' to punish your dog in any way, shape or form. We should question everything, even when advice comes from 'experts'.

In my behaviour practice, we frequently see dogs that perform highly complex and enduring behaviours, which they have learned inadvertently. If these weren't problematic, they would be miraculous. For example, if I were to ask the average pet owner to train their dog to run to them at high speed each and every time they gave a cue, no matter what the dog was doing at the time, they would be likely to have trouble. If I were to tell them that they could only practise it six times first, the task would become impossible. This is ironic considering that the doorbell performs this feat with 100 per cent reliability after an average of only four repetitions.

Of course, it is easy to see why such behaviour gets established so quickly. Despite the fact that the doorbell means nothing to a dog on the first occasion he hears it, it quickly becomes paired in his mind with all the excitement that follows, and therefore becomes a cue for dashing to the door.

As we established with Zeus, the Afghan Hound, dogs are so good at pairing events and consequences that they can easily recognise the start of even long sequences or patterns which will eventually result in a reward (or, for that matter, a punishment). My own dogs do a remarkable impression of being telepathic, seemingly knowing when we are expecting visitors – an event they love. Despite my attempts to behave as normal, literally hours before a visit my Collie-cross will run to the window that overlooks the drive and bark in anticipation, just in case. Clearly, something about my demeanour triggers her behaviour, and then the rest of the 'visitors are coming sequence' confirms that she is right. In my case, this usually involves vacuuming, tidying and making sure there are no dog hairs in the sandwiches. (For this reason alone, it's worth giving me several hours' notice if you'd like to drop by for a cup of tea.)

Roxy was a sleek and shiny English Pointer. At three years old she was in the peak of condition, and met me at the door of her owners' home with nose nudges and her tail whipping hard against my legs. I followed her into the living room and chatted to her owners, Alan and Annie, about Roxy's history.

Roxy had been bought from reputable breeders as a potential show dog. When she arrived at their home, Alan and Annie had been careful with her socialisation, making sure they took her out and about to meet lots of people in different places during her first weeks and months. It had surely paid off, for they described her as an outgoing and affectionate dog, who loved contact and attention from humans. Roxy was also good with other dogs, although she was rarely let off lead to run or play, as her owners were worried that she might hurt herself and damage her show career.

Roxy was on the go most of the time while we were talking, bringing toys and nudging my hands to elicit attention. Although demanding, she wasn't pushy and I felt comfortable in her presence and safe making contact with her.

Alan and Annie were keen to tell me her story.

'It all started about a year ago,' said Annie.

Alan nodded in agreement. 'Yes, we were sitting watching some TV programme – one of the soaps Annie likes, probably – when Roxy started staring into that corner.' He indicated a corner of the white painted ceiling. It appeared unremarkable to me.

'That was it. From that point onwards, she wouldn't stop staring. We were convinced she could see something we couldn't.' Alan looked slightly sheepish, as if he was trying to gauge my reaction. 'In fact, it got so bad that we thought she could see something, well, supernatural.'

There was a pause. He went on. 'It started in that corner, then she would stare at the other corner and a spot over there near the lamp. Sometimes she'd even growl. That would really spook us. She'd just stare and stare at the ceiling and growl as if something was there.'

I asked when they saw Roxy do this, and what her body language was like was during the behaviour. This is important because it's not uncommon for dogs to stare at shadows and reflections, and becoming obsessed with them can even be a symptom of a clinical problem. In our practice we have video footage of various breeds showing behavioural symptoms of clinical problems, such as low-level seizure activity. There is no great dramatic collapse or fitting in these dogs – just odd behaviours such as non-recognition of their owners, obsessive licking, spinning, fly snapping and repetitive pouncing. Such problems tend to be difficult for a vet to diagnose as the symptoms aren't seen in the veterinary surgery, and they can also be indicative of many other disorders, some serious and some not.

'Initially, she would only do it in the evenings, but then she'd do it pretty much any time we'd been in here for a bit. Sometimes she just stares, but on other occasions she'll do the full "pointing" action, with one front paw raised and her whole body quivering with attention. She never does it in front of visitors, though,' said Annie. 'Except my mum once – she was staying here and she saw her doing the staring thing. I think it freaked her out a bit.'

I scratched Roxy under the chin and she leaned against my leg in a friendly fashion. It certainly didn't sound as though Roxy was in distress during these spectre-staring episodes, although on-lookers might be.

'And you're pretty sure that she doesn't do this behaviour when you're not here?' I asked. When we spoke on the phone, I'd suggested that they leave a video camera running to film Roxy while she was on her own.

'It was quite funny to see what she does when we're not here,' Alan said. 'She had a play with her toys, looked out the window by jumping onto the arm of the sofa, and generally chilled out. She didn't seem at all worried, and she didn't look at the ceiling once.'

This was great information. Dogs suffering from clinical problems don't just switch them on and off depending on whether we are in the room.

Alan looked slightly embarrassed. 'Annie was beginning to think that the house was haunted and that we should have it exorcised.'

'You did too!' exclaimed Annie, looking indignant.

'Well, I don't really believe in all that stuff, but it's true that dogs can see and hear things that we can't, isn't it?' he asked.

Keen to leave the debate about things that go bump in the night for another time, I thought I'd try a little practical experiment.

'OK,' I said, 'we're all going to ignore Roxy and see what she does. No eye contact, no talking to her, no touching.' I folded my arms and looked away from her. The dog practically did a double take, and walked round to the other side of the sofa to see if she could attract my eye contact from there. I turned my head away again, making it clear that I was not going to engage. Roxy turned and did a big bounce onto a toy – front end rearing up and then pouncing like a cat on a mouse. In a wild animal this is called a 'fore-foot stab', and it's such a classic, graceful canine behaviour that it was hard not to watch. Alan guffawed.

'Try not to look at her, no matter how cute,' I warned.

Bemused, Roxy took the toy over to Alan. Following my lead, he turned his face away and ignored her. She tried again, letting the toy fall then picking it up and throwing it at Alan. Finding herself ignored still, she walked over to Annie, put her front paws up on the chair she was sitting in and tried to get on her lap. I could tell that it took all of Annie's restraint not to laugh and give her a cuddle. This dog's attention-seeking behaviours were pretty endearing.

They were also well practised and clearly successful – which is fair enough. After all, if you turned on all your cutest charms, wouldn't you expect your husband or wife to notice? If you're in a social relationship it's natural and rewarding to engage your family member in mutual communication, attention and response.

Roxy was clearly good at getting attention, but if her odd behaviour was linked to this, then she wasn't showing it now.

I asked the couple what they had tried so far to help resolve the behaviour.

'Oh, everything,' they answered. 'We've said "Uh, uh," to tell her off. We've got up and investigated the area she's staring at to try and reassure her. We've even thrown a tin can full of pebbles to startle her – we saw it on TV – but she just stepped round it.'

English Pointers are renowned for being a good-natured breed that love human company. They also love to work and, as the name suggests, they are bred to 'point' at birds or other game in the field for hunters to then shoot. Clearly, Roxy was using some of this strong breed instinct to point at things in the home, but without knowing what was triggering it, we were going to struggle to help her.

Alan reached down to the side of the TV and picked up a DVD case. 'Do you want to see her in action?' he said.

'Oh, absolutely.' I was surprised because the couple had told me they hadn't been able to take any footage of the actual behaviour.

'When did you manage to get that?' I asked.

'Yesterday evening,' said Annie. 'We were sitting here as usual and she started. I managed to grab the camera Alan's dad had lent us and we got a bit on film.'

Alan opened the case and slid the DVD from the box. He reached over to grab the remote control. Out of the corner of my eye, I saw a tiny shift in Roxy's expression. Something changed, as if she had tensed just a little. She walked round the back of the sofa and I turned my body to watch as she re-emerged on the other side. Alan popped the DVD into the sliding drawer of the player. He turned on the TV. Roxy positioned herself in front of me, body facing towards her favourite corner of the ceiling, head swivelling between her owners and the wall.

'That's interesting,' I said.

Alan and Annie followed my gaze and looked at Roxy.

'Alan! She's going to do it!' exclaimed Annie. Alan hit pause button on the DVD and stopped still. We all looked at Roxy. She looked back.

'Let's ignore her again,' I suggested, thinking that staring at the dog might be intimidating. Roxy seemed to relax a bit and came over to sit next to Annie.

'Ah, well. Looks like the TV version will have to do,' I said, as clearly the live show was over before it had begun.

Alan pressed the remote and the DVD came to life. Our attention now focused back on the TV, it was easy to miss Roxy heading over to her 'pointing spot' once again.

'Look,' I said quietly. 'She's getting ready again. Let's leave the DVD on for a second and see what happens. Don't look at her.'

Roxy started to stare at the wall. While her other self was doing exactly the same thing on the TV screen in front of me, there was a real-life dog gearing herself up for a full 'point' at the corner of the ceiling.

'Do you think she's being triggered by seeing herself do it on TV?' Annie asked. 'She's never done it in front of anyone else before.'

'I'm not sure,' I said. 'Let's turn the sound off and see what happens.'

Alan killed the sound, and Roxy flicked her head round.

'Let's just watch the screen for a minute,' I said. We kept our faces turned towards the TV. The bizarre behaviour on the screen continued to be mirrored in the room. Roxy was in full-on English Pointer mode now, front paw raised, eyes fixed and focused. I could see why her owners found it unnerving.

I asked Alan if he could change the channel so that we were watching TV, not the DVD. Roxy changed her position, and started to stare at a point above the lamp. 'She's doing it,' Annie said.

Indeed she was. Roxy continued to stare at the wall, and even gave a little growl. I asked Alan to switch the TV off. Within seconds, Roxy shook herself, resumed her normal stance, and headed over to Annie to see if she could get a tickle.

'It can't have been the film that set her off,' said Annie. 'Maybe it's the TV.'

'Is it giving out some kind of sound that she can hear and we can't, like white noise?' asked Alan.

A nice thought perhaps, but it seemed to me that the answer was more simple than that. In the next half-hour, we repeatedly switched on the TV and switched it off again. We left it on standby and turned it off at the mains. We watched all the different channels and put the DVD back on, just to test our theory.

Roxy's ghost-hunting was triggered by her owners watching TV. It didn't matter what they were watching. If their attention was tuned to the screen she would resort to her staring antics within minutes. How this clever dog had hit on the idea of staring at the ceiling in the first place would, of course, remain a mystery. Perhaps she had glanced up there one evening while Annie was engrossed in a programme and had been rewarded by her owner looking over at her. Maybe there had really been something there on the first occasion – such as a shadow or a fly or a spider – which had caught her attention, and then this had become a learned behaviour through the simple mechanism of reward and repeat. We would never know. What we could be sure of was that Roxy was telling us she was in need of some constructive hobbies.

First and foremost, however, we needed to give Alan and Annie some control back over what was happening in the evenings. It was clear that Roxy had a great quality of life, but her imaginative use of the down-time when her owners watched TV was not going to disappear without some intervention, simply because her ability to engage in a breed-specific motor pattern was so rewarding in itself. In other words, if you are a Pointer, pointing is fun. In such cases, attempts at 'interrupting' the behaviour with relatively mild punishments such as shouting, or throwing rattle cans full of coins or pebbles, are likely to have zero impact. Most dogs, even sensitive ones such as the English Pointer, will simply learn to avoid the unpleasant event – as Roxy had by side-stepping the missile – in order to continue with the behaviour that is giving them pleasure.

The first part of Roxy's behaviour programme was therefore going to start with Annie learning how to use the TV recording facility. Why? Because she wasn't going to be able to watch much TV in the next few days while she was retraining Roxy. Instead, each and every time Roxy even thought about staring at the ceiling, she was going to quietly get up, catch hold of a short trailing line we had attached to Roxy's collar and take her calmly into the kitchen for a two-minute 'time out period'. With my help, Annie and Alan practised this, and within a few trials, Roxy was already looking less sure of her previous strategy. This was going to take some consistency and practice, but I felt hopeful that Roxy would catch on quickly.

However, it's one thing interrupting a behaviour and quite another making sure that we are addressing the root cause. Roxy's need to do some 'pointing' was obvious, but as two hard-working urban dwellers, it simply wasn't going to be possible for Alan and Annie to train Roxy as a gundog in the field. Instead, we needed to come up with some strategies that would enable her to use her incredible Pointer nose, so that she could point to things she had found by scent rather than by a flight of fancy.

I have always enjoyed teaching my own dogs to play at 'drug detecting', using a sachet of dried catmint hidden amongst the soft furnishings of my living room as the 'drug', and I thought that a similar game might appeal. In the style of all good detection dogs, the 'drug' must be found but not touched, and so 'pointing' at the spot where the article was hidden seemed ideal for Roxy. We started slowly, teaching Roxy that if she sniffed at the catmint sachet – actually a cat toy, but we didn't tell her that – she would receive a click and a treat. We then began to 'hide' the sachet; first in easy places, such as by a chair leg or under the edge of the long lounge curtains, and then in increasingly more challenging spots, such as under cushions. If she found it by sniffing and then indicated it was there by looking at it, she got her click and treat. Scrabbling at it with her paws or trying to pick it up got no reward.

Two weeks down the line, Alan and Annie called to give me an update. Roxy had all but given up looking at the ceiling when they turned the TV on. In fact, they'd only had to take her out of the room once in the last five days, and were very pleased with the result. I asked how the games were going.

'Roxy absolutely loves the hunting games,' said Annie. 'In fact, she's got so good at it that we can

hide the catmint in the garden, even the park, and she'll search for ages to find it. We showed off her "drug detection" skills to a friend who came round by secretly hiding it in her handbag when she wasn't looking.' There was a pause. 'There's only one thing though.' I could hear Alan laughing in the background. 'How do we stop her from pointing at the cupboard where the sachet is kept?'

TOP TIPS FOR HARNESSING YOUR DOG'S LEARNING ABILITIES

» Dogs love learning new things. Given the right motivation and with kind methods, you can teach your dog to perform all sorts of amazing behaviours. My dogs sneeze and yawn on cue. Not useful, it's true, but great fun.

» 'Sit', 'down' and 'come when called' are essential basics for all dogs, but don't be tied to traditional training. Tricks such as spinning, walking backwards and 'speaking' on command are fun and can be incorporated into obedience exercises too.

» If your dog is behaving in a way you don't like, the first question you need to ask is 'How is this being rewarded?' Note that this is a far cry from asking 'How can I stop this?'

» We should bear in mind Milgram's psychological experiments of the 1960s. All too often 'experts', whether on TV, in a book or in the park, tell us we should use confrontation or punishment with our dogs – and it can be surprisingly difficult to resist. The answer is to question everything.

» Be imaginative when it comes to allowing your dog to express his natural instincts. I don't have a flock of sheep for my Collie-cross to herd, but she gets the same thrill from herding giant gym balls like those used for Pilates, in a new dog sport known as Treibball.

» Use clicker training (see page 22) to help teach a new behaviour or skill. It's popular because it works.

....

The fact that dog behaviour is closely linked with human behaviour will always make it fascinating to us. Behaviour work may seem glamorous, mysterious even, and, as a result, it could be said that training has become the 'poor relation' of behaviour modification. While I believe that behavioural understanding is the vital foundation for good training, I also think that good training is the foundation of most behavioural solutions! One cannot exist without the other, and although I like to

explore many different psychological angles when approaching a case of problem behaviour, I nearly always turn to basic training as my starting point for achieving change. It can be an elegant, fun and fast way to offer solutions that will help owners to manage problems and sometimes even resolve them.

Contrary to popular belief, training is not manipulation or coercion. Training, in my view, is about teaching, and that means having an excellent understanding of the animal's emotional state as well as its learning style.

Imagine someone training a cat. They would surely realise that they need to motivate the cat to want to be in the same room as them in the first place – and not just in close proximity, but confident in their presence. After all, it's pretty hard to train a cat if it's hiding under the sofa and refusing to come out. This means that a good trainer would be likely to spend some time allowing the cat to become familiar with them, to feel comfortable in their surroundings and to know that they were safe. These goals might be managed in practical ways – for example, by sitting down and waiting for the cat to come near rather than attempting to approach the cat to say hello, or by feeding the cat from their hand – but their aims are emotional. If the cat doesn't feel safe, he'll want to avoid the trainer and is unlikely to learn anything, apart from how to escape. Once the cat's emotional welfare has been established – and it may take some time – it needs to be maintained. The training may well start, but do you think the cat trainer is going to

risk wasting all the time and patience they've invested by punishing the cat when it makes a mistake? I think not!

Training animals other than dogs teaches us many things. In my time, I have been lucky enough to train rats, hamsters, cats, pigs and even chickens, all using positive reinforcement methods and, most especially, clicker training.

This behavioural/emotional feedback loop may seem simple, and to a certain extent it is, but it is also incredibly powerful in forging bonds between human and dog, in creating a symbiotic relationship between two very different species and in helping to form a very special team.

13

Training in a foreign language

ET come home

Case history: **ET the scent-addicted Puggle**

It was a meeting with a Beagle-cross that got me thinking about learning styles – in dogs, as well as people.

ET the Puggle snuggled on my lap and enjoyed having his ears scratched while his owner, Jo, brought cups of tea and told me his story.

ET had come from a breeder at eight weeks. A deliberate breeding between a Pug and a Beagle, his resulting 'alien' good looks had earned him his name. A fun and energetic puppy, sociable and friendly, and happy to run around the garden with the kids, he had charmed the whole family, as well as neighbours and friends.

Right from the start, however, the hound part of ET's make-up had made itself known. On occasions, he would be completely and utterly governed by his nose. Without warning, the little dog would suddenly stick his muzzle in the air, turn on the spot and follow, 'Bisto-kid' style, whichever way the scent dictated. No matter that it meant he had to scramble up and over the four-foot garden fence, or run across roads, fields, or parkland; once ET had that smell in his nostrils, nothing would stop him.

Jo quickly realised that ET was not like other puppies. When out on a walk, he would be trotting at her side one second, then take off at high speed the next. Even as a youngster, he would run at full pace, nose down, following a scent for miles and miles with her trailing behind, calling desperately. She had opted for using a long line after the third occasion on which he crossed a busy road with her in hot pursuit, oblivious to her shouts.

Poor Jo was determined to resolve the issue and in the last year she had tried everything in her power to train ET to listen to her more, but to no avail. She had attended puppy classes, dog-training classes and had one-to-one sessions with an instructor. She had read countless books, had extended the height of the garden fence and had even bought a 'remote training collar' designed to give the dog a blast of compressed air to make him stop and listen – all without effect.

Once ET's nose was switched on, his ears shut down, and it was making ordinary family life with their dog almost impossible.

ET set my thoughts racing. In people, there are lots of clues about the 'representation system' someone is using. For example, a highly visual person might say 'I see'. Auditory types may use phrases such as 'I hear you' or 'Loud and clear!' People who are highly kinaesthetic, or tactile, may say 'Keep in touch' or 'I feel that's right'. Recognising how someone represents the world means we can offer them information in a way that best suits their learning style. Ignoring their learning style creates frustration, challenge and dissonance. Frankly, it's poor training.

Some 25 years ago I was introduced to a psychological method known as Neuro-Linguistic Programming, or NLP. Despite having the most off-putting name, NLP turned out to be quite an eye-opener for me. I've enjoyed studying it ever since, and qualified as a Master Practitioner some years ago. Of course, my first love is animal behaviour but nearly all dogs and cats come attached to an owner, and that means I need to understand human psychology as well. After all, if I can't communicate with the human end of the lead, the dog doesn't have a hope!

As a trainer, instructor and lecturer, I am also a teacher and a coach. (This, much to the wry delight of my teacher mother who clearly hoped some of her expertise had rubbed off on me.) However, gone are the days of the 'mug and jug' approach to learning, in which the teacher was expected to 'pour' from the jug of knowledge into the waiting student 'vessel'. Instead, new approaches look at the way humans learn: how we experience the world, and how this impacts on our understanding. NLP offers some unique and fascinating insights into how human beings function, how we think and feel and how we can project our inner-most thoughts and experiences to create our own 'reality' – all of which impact on our behaviour.

When I was at junior school, I struggled with maths. Sciences weren't a problem, but maths seemed to me to be dry and barren – somehow separate from real life. Whilst it was likely that I would scrape through my senior school maths exams at a basic level, I didn't enjoy the subject. Maths lessons were pretty torturous for me (and for the teacher too, no doubt). A year went by and I was assigned a different teacher with a completely different approach. Maths was taught using diagrams, visual representations and applications to real life. Suddenly all those mysterious concepts and equations meant something to me. It was like being taught a whole different subject, and I loved it.

Looking back now (an important choice of words!), I realise with clarity what caused the change. Human thoughts and experience are structured in terms of our senses. When we think, or process information internally, we 're-present' the information in terms of the sensory systems that are our contact with the outside world – and the way we do this is very individual.

Imagine a bonfire. You may be one of the many people who immediately form an image in their mind of a bonfire, complete with colour, flickering flames and intricate detail. It's like watching a movie image of the scene. Alternatively, you might be more inclined to hear the crackle of the flames, the snap of the twigs as they are consumed by fire, the hiss of damp branches in the heat. Primarily auditory people concentrate on sounds, words and voices. Kinaesthetic people, on the other hand, are those who access information primarily via touch and feelings. They may imagine the warmth of the flames, or they may respond emotionally to the feelings that are conjured up by bonfire nights of their childhood, or other experiences.

In a nutshell, people tend to access information about the world around them in five main ways:

Visual – Seeing
Auditory – Hearing
Kinaesthetic – Feeling
Olfactory – Smell
Gustatory – Taste

Although we all use a combination of these senses, as someone who is highly visual, I access the world and 're-present' it primarily in pictures. I remember scenes, details and even words in a visual way, with colour, brightness and clarity being important to the images. While I do pay attention to some kinaesthetic and auditory inputs, place me in front of a good TV drama and I won't even hear you come in and offer me a cup of tea – I'll only notice you when you step into my line of vision.

This principle of differing learning styles has major implications for those taking training classes, doing behaviour work, or simply attempting to communicate an idea to another person. For example, trying to teach a visually oriented person how to get their dog to sit simply by telling them is going to be hard – unless you paint pictures with your words. Equally, asking a primarily kinaesthetic person to ignore their dog and not touch it will be downright impossible!

So, what if dogs were telling us which representation systems they preferred and we were ignoring them? What if the principles of NLP could be applied to dogs' learning styles, despite the fact that they can't use language? What if this affected whether training was a success or a failure?

From my own observations and experiences with dogs over the last twenty years, it's apparent to me that different breeds or types of dog learn in different ways, and these seem remarkably similar to those that humans experience – with some notable additions due to the dog's enhanced senses of smell and taste.

Think of a 'visual' breed. Although it might be an odd request, most people can easily find an example. How about a Lurcher or an Afghan Hound? Perhaps it's no surprise that dogs in this group have been bred to hunt by sight, and have prey drives that are triggered by movement. Some even have eye shape and placement that reflect how important this sense is to them, as it allows maximum peripheral vision. Collies, and other herders too, appear to be primarily visually oriented, although they perhaps have more of an auditory balance (after all, the herding needs to be controlled by a human whistling or calling commands).

Could it be that terriers are the 'auditory' group of the dog world? Highly tuned to listen for tiny noises that might indicate hidden quarry underground, they need to be sensitive to sounds of all kinds. They also enjoy making noise themselves – earning some a reputation for being yappy.

Gundogs could be regarded as kinaesthetic. Although they are great at finding items by scent, they seem to gain a great deal of information through touch, almost completely to the exclusion of auditory signals (unless specifically trained otherwise). These are the dogs that most love to be touched, and find crashing through undergrowth, jumping up at people or being petted highly rewarding. Bored gundogs tend to choose chewing over barking. Maybe a heightened sense of touch is a vital part of needing and wanting to pick things up – a skill that is highly evolved in many of the gundog breeds.

Of course, in comparison to dogs, humans are woefully lacking in olfactory sensitivity. Clearly, many of the hounds live in a world of scent and learn about the world through an olfactory representation system. Perhaps they make 'scent pictures' in their heads? It's little wonder that although all dogs can be trained to follow scent, it's the Bloodhounds, Beagles and Foxhounds that really excel. Those dogs have been bred for centuries to be able to detect and follow even the tiniest disruption to vegetation while on a scent trail. It's hardly surprising that they ignore auditory input when their brains are fully engaged with sniffing. Their focus is so great that they simply don't hear, or feel, much else at all.

Just as in humans, having a primary learning style does not – and should not – mean that a dog can't take in information in other ways. I wonder whether the ability to use a combination of learning styles makes dogs' behaviour more rounded and more stable? Certainly, many of the dogs I see for behavioural problems seem to lack sensory 'balance', and I wonder if this contributes in some way to their behavioural issues? Many I see are obsessed with looking at the object of their anxieties, are overly sensitive to sound or touch, or seem to be addicted to physical contact.

In order to investigate my theory, I reviewed three years' worth of case histories of dogs that I had seen for behaviour problems, looking for a strong propensity towards one of the accessing styles rather than a balance between all five. While this was a personal survey based on observation rather than a strictly scientific study, it threw up some fascinating results. I discovered that there were several main areas of behavioural difficulty in which a single-minded focus on one sensory system seemed to have an impact – and although this had a slight breed bias, the differences appeared to be highly individual.

In more than 75 per cent of genuine separation anxiety cases that I had seen, the dog in question seemed to be obsessed with either touching or seeing his or her owner. These dogs wanted to follow their owners around the house, come into the bathroom with them and lie next to them on the sofa. However, they weren't content just to lie close by; they needed to be draped over their owner's feet or, even better, lap, in order to feel the reassurance of the 'fix'. If the owner moved their foot away by even one inch, these dogs would subtly shift position to be in contact once again.

Visually 'addicted' dogs tended to be determined to watch one or more members of the family. Prime spots for such 'visual fodder' were at the junction of several rooms: at the top of the stairs or the threshold of a doorway. Did this indicate the dog was being 'dominant' in any way? Absolutely not. He'd simply found the best watching spot in the house.

Nervous dogs, too, seemed to have an above-average need for their owner's physical presence. They typically showed a kinaesthetic need to be reassured much of the time by touching their owners, nudging their hands, hiding behind them, or by having a physical connection via their collar and lead. While we think of pulling on the leash as being bad behaviour, it's possible that such dogs are gaining reassurance from the physical feedback of constant pressure round their necks.

Sadly, for some dogs, following a single-minded way of accessing information may look to their owner like simple disobedience. A hound with his nose to the floor probably doesn't physically hear or feel any other input in the moment, and this leads to human frustration. The temptation is to try and make the dog listen, to make it feel, but there is little chance that it can do so while its brain is otherwise engaged.

ET sighed in pleasure next to me on the sofa. He clearly had no problem relating to people when he was relaxed. Jo explained that she had tried everything to get ET's attention, but when his nose was on the floor she could be standing right next to him and he wouldn't even look up at her. It was as if she simply didn't exist in his olfactory world.

Clearly we were going to have to modify our training methods to suit ET, not the other way round. Of course, this presents its own challenges because it can transpire that the human end of the lead also has a strong predilection for one learning style over another – which may not be the same as the dog's.

For years, owners have tried to train dogs using the same 'one size fits all' methods. The reality is that any method that relies on physical contact is going to be ineffective for dogs that are simply not very kinaesthetic. Even for those that are, the effect of human touch may be exciting rather than educational. Verbal cues or commands aren't going to help much either unless your dog is highly auditory. For example, there seems little point in trying to teach a highly visual breed, such as a Greyhound, how to lie down on cue by simply repeating a spoken command. Combine that with trying to manoeuvre his front legs into position by pulling them and if he's sensible, he'll simply avoid you whenever he can. You are not talking his language!

There was little doubt that ET lived on a different planet from Jo and the rest of her family. While humans enjoy the sights and sounds of the countryside, houndy pups like ET are completely engrossed by the smells that surround them – both in the air and on the ground. Once his nose was engaged, it was too late to try to get ET to switch information-accessing strategy and swap to vision or hearing – it was all about the scent.

In this country, scent work has been largely confined to working dogs, such as police dogs, or sniffer dogs at ports looking for narcotics and explosives. There are also some charities that train dogs to detect certain scents for assistance work; for example, they can learn to predict the onset of an epileptic seizure, sniff out cancer, or alert diabetics to changes in their blood-sugar levels.

While some enthusiastic pet owners use scent work as a part of another competitive discipline, such as obedience or working trials, little has been available for the average pet dog owner who simply wants to have fun harnessing the power of their dog's scenting abilities.

Our task was to change all that and think like a Puggle! During the rest of the session with Jo, we made it a mission to harness ET's scenting skills by teaching him to follow a track. Done correctly, tracking helps dog and owner to learn trust in one another and to work together as a team. While it was clear that ET already knew how to track like a pro, Jo needed a few lessons to make sure that her dog was following the trail of scent she had determined and not some random one. With this aim, we decided to start by using a portion of ET's food and his favourite toy.

Jo and I packed some kit and put ET in the car. We travelled to a big park not far from Jo's home and chose an area that is rarely used by other walkers, so the ground would be fresh. ET stuck his nose out of the crack in the car window and sucked in air and scent – but it wasn't his turn yet. With ET watching from the car, Jo and I set out our plan.

Carefully, we stuck a pole in the ground to mark the start of our track. Basic track laying is not complicated, but there are a few things you can do to ease your dog into the new routine. The pole acts as a visual cue to mark the start for the human handler but over time, it also becomes associated with the start of a scent trail for the dog, and means 'Put your nose down now'.

Jo took exactly fifteen steps in a straight line away from the pole. She then stopped, placed ET's beloved tuggie toy on the track and covered it over with some long grass. After another ten paces, she placed a pot of food on the track (one with a tight-fitting Puggle-proof lid). She walked another five paces, stopped, then retraced her steps all the way back to the start. It's important to continue along the track a little way after the final 'goal' to prevent the dog getting into the habit of stopping once he's made a find.

By the time we got ET out of the car, the little dog was already anticipating something exciting. He looked up at Jo, and she commented on the fact that he seemed to be asking her for information, not just gathering it on his own. ET's nose quivered like a prehensile limb, and he pulled against the harness towards the start of the track. Jo and I had agreed on a pre-arranged sequence. Once at the pole, she said the dog's name. Unsurprisingly, he ignored her, so she took two steps back, away from the pole. The little dog looked up at her in surprise. That was her opportunity. 'Good,' she said, quickly moving forward. 'Track on.' Jo moved her hand down to the ground to indicate the start of the scent trail she had laid.

ET didn't need to be told twice. His nose went down to the ground and practically sucked up the grass as he moved along the trail. Jo praised him for going the right way and allowed him some slack in the line she held. After only a few seconds, ET had nearly bumped his nose on the tug toy hidden in the grass, but didn't find it. He started to explore the area around the toy, his sniffing growing in intensity.

If dogs are so good at scent detection, I hear you ask, why is it that they can search for ages for a ball or toy in the grass that you and I can plainly see? Although this seems like a strange irony, it is because first and foremost dogs look with their noses and scent can be a tricky customer. For example, we might see that a toy is lurking in the grass right next to the garden fence, but the dog could be hunting for it three feet to the right – which is where the scent 'pool' has been blown by the breeze, buffered by the fence.

Most humans don't even consider scent 'pictures', unless they are smuggling contraband for a living. They vary greatly depending on temperature, air movements, the size and type of the environment, and the size and type of object the dog is searching for, as well as how long the object has been there and what other disruption there has been to the trail.

Now, while it's very satisfying to watch your dog following a track, it's important for pet dogs that you interact with them along the way. This interaction is what helps the dog realise you are an ally, a useful member of the tracking team with relevant information and skills that will help with the task in hand. However, this is quite different from interfering! Jo allowed ET to cast about for a few seconds, searching for the toy, then she stepped towards him and with a slow, assured hand movement – as if showing someone to their seat in a theatre – she guided ET's attention back to the point where the toy was hidden. ET glanced at Jo – the first time he had even noticed her presence since the track had started – then went to where her hand indicated. He nudged the toy and unearthed it from the grass, flinging it in the air in a show of surprise and delight that made Jo and me laugh out loud.

As instructed, Jo stepped off the track slightly, took the end of the tuggie toy and had a wild and wonderful game with ET to build some belief in the little dog that paying attention to your tracking team-mate is worthwhile. Once the game had started to subside, Jo pocketed the toy then quickly encouraged ET to put his nose back on the track. ET found himself in a quandary: the drive to follow the track once again was strong, but so was the idea of having another game with the toy that he knew was in Jo's pocket.

Jo acted in her role as team player. 'Track on,' she told ET, and directed his attention back to the track. It only took ET a few more seconds to discover the pot of food. Seeing that ET had found it, Jo moved towards him and again, ET had to wait for his team-mate to come and release the reward – by taking the lid off the pot. Once he had eaten the food, ET's lead was clipped to his ordinary collar and we trotted quickly back to the car.

Over the next seven days, Jo and her kids went out and laid a track for ET at least once a day: sometimes in the garden first thing in the morning and on other occasions in the park or local woodland. As instructed, at first they kept the tracks short and sweet, but they were soon able to extend the trail in length and then in complexity, by making right-angle turns en route. ET was well and truly hooked. Subtle changes had started to occur in his general behaviour at home. The family noticed that he seemed calmer and less likely to charge into the garden, ignoring all human protests, when the back door was opened. Instead, he would wait to see if Jo or one of the children was going to come and put on his harness to do some tracking.

Out on walks, ET would still want to put his nose down to vacuum up the smells of the great outdoors. However, Jo had started to incorporate other 'finds' into her tracks so ET never knew what he might be searching for next. Her children made her a collection of odd items: a piece of hosepipe about eight inches long, a washing-up sponge cut in half, a piece of knotted rope, a matchbox. She carried them in her coat pocket so that they would be covered in her scent and ready to be placed on a track – or dropped as she walked along. She got good results if she sent ET back the way they had come to search for them. Once he found them, he would often pick them up to

bring to Jo, and she would instantly swap them for ET's favourite tuggie toy and have a good game.

Three weeks on and Jo made a break-through. ET was still being walked on his long line, which kept everybody safe and made sure that walks and tracking were stress-free. However, Jo started to notice that ET was beginning to keep an eye on her. Instead of walking with his head down, nose glued to the ground, he would glance up from time to time to see where she was. Turning her body towards him now seemed to catch his attention – presumably in the hope that she would send him off on a mission to find secretly hidden matchboxes and pieces of hosepipe on the path they had just taken. Sensibly, Jo started to reward these split seconds of genius and gave ET treats, praise and games with his tug toy when he looked up at her.

After six weeks, Jo's commitment to ET's training was really paying off. She still needed to focus on him when out on walks, and would keep him on the long line when she was short of time or just wanted to relax after work. However, she could let ET off the lead to go and search for articles she had covertly dropped, sending him back along the route they had taken and even directing him into bushes and trees at the sides of the path to hunt for them. The family found this strategy so successful that they decided to keep the food pot for garden use alone, and this proved useful in keeping ET's interest within the boundary of their fence.

There's little doubt that ET now has more freedom than he used to and enjoys his off-lead sessions in the park and the woods. While Jo is realistic about the strength of ET's natural drive and his single-minded focus on the olfactory world, her efforts and training have helped to build a bond with her dog that will last a lifetime.

Jo rang me to tell me about one further bonus. During a woodland walk when ET was searching for items that she had hidden in the undergrowth, Jo noticed that he was struggling to pick something up in his mouth. All the articles she had been using were deliberately chosen because they were safe and easy to carry for a little dog whose muzzle is definitely more Pug-shaped than Beagle. ET persisted and as Jo came forward to help him, she was amazed to see the glint of metal between his teeth. ET emerged from the undergrowth just at the edge of the path, wagged his tail and moved towards her. Jo couldn't believe her eyes as he dropped her car keys at her feet!

TOP TIPS FOR TRAINING IN LINE WITH YOUR DOG'S LEARNING STYLE

- Assess your dog's learning style by considering his breed, type and individual characteristics. Can he copy what other dogs do just by watching? Is he tuned in to the smallest sound? Does he need and want to touch you all the time?

- Think about your own learning style as this will inevitably affect the way you approach training. A very similar learning style to your dog (for example, you are both highly visual) makes training relatively easy. Very different, and you might need to adapt in order to accommodate your dog's needs.

- Beware of imposing your primary representation system on your dog. Dogs that hide behind their owners' legs or cling to them when they are worried may be reflecting their owners' kinaesthetic needs rather than their own, especially if they have been reinforced for that behaviour.

» If your dog's learning style is visual, it makes sense to concentrate on teaching hand signals and visual cues. You can pair these with verbal commands, but be aware that they will be less salient for your dog.

» If your dog is primarily auditory, verbal cues will be easier for him to learn and recognise. Tone of voice will have more of an impact too, so be careful how you express yourself so as not to load your cues with emotion.

» The types of reward that your dog likes will depend to some extent on his or her learning style. Visual dogs love chase games, kinaesthetic ones like touch, auditory ones find squeaky toys (and being allowed to bark) reinforcing. Gustatory dogs like food!

» Scent work is said to be difficult to teach and only for those who have a working gundog or a police dog, but nothing could be further from the truth. All dogs, no matter what the breed, have a great sense of smell; it just needs to be channelled in the right way.

Overall, a well-balanced dog probably has well-balanced learning styles. If your dog relies on one learning style to the exclusion of all others, try increasing his awareness in the other areas by developing skills that relate to those sensory inputs. For example, teach your dog to track if he is primarily visual; train him to focus and watch you if he is highly auditory – and use his natural learning style as a means of doing so.

Conclusion

A dog by my side

Those of us lucky enough to share our lives with canine companions need little persuasion to convince us that our domestic dogs are sentient, emotional and intelligent beings. They are not, thankfully, little people in disguise; nor are they merely a 'watered-down' wolf, but a species all their own, with their own needs, wants and abilities.

New evidence suggests that the co-evolution of man and dog is no coincidence – indeed, it may well have been co-dependent. For most dog owners, it is not difficult to believe that man has been successful by virtue of this creature – one so attuned to human social communication that he enabled us to hunt better, to live more safely and to protect our resources. The dog is the 'perfect peer' when it comes to social co-operation, attachment and communication and, what is more, these animals seem to understand us.

Most people now accept that humans evolved from apes. However, despite the fact that many apes live in social groups, they are often aggressive, competitive and selfish, especially when resources are in short supply. If you're watching the TV news it's possible to believe that we are not much different, but this is factually untrue. Humans live in much closer proximity to each other than apes ever have. Put a hundred chimpanzees on a crowded tube train with nothing but each other for company and a fight to the death would break out within minutes – if not seconds. As humans, however, every morning and evening we politely shuffle along to let just one more person inside the carriage, and although we may not enjoy the experience of being packed into a hot, confined space with complete strangers, true aggression is incredibly rare.

Social co-operation may have evolved in humans essentially because, just like dogs, we too are eternally 'childlike'. It is interesting that adult humans are remarkably similar in development to the juvenile stages of some of the apes. We fail to develop hair all over our bodies. We have rounded heads and big eyes. We enjoy close physical contact with family members. Is it possible that in order to be social, many of our ancestors' aggressive, competitive, adult characteristics got arrested in development? As 'big kids' we can show all the sociability we like, from playing games with each other (who said football was childish?) to hugging (about as far away from demonstrating social dominance as you can get).

New views on human evolution suggest that man's ability to be social may be in part due to being 'eternally infant', just like dogs, allowing rich

communication and co-operation. Man is not ape. Dog is not wolf. The child and puppy need each other just as much today as they ever did – and it's up to us to make sure that we both survive and thrive.

Bibliography

Gavin de Becker, *The Gift of Fear: Survival Signals that Protect us from Violence*, Little Brown & Co. (1997)

Clive Bromhall, *The Eternal Child*, Ebury Press (2003)

Ray and Lorna Coppinger, *Dogs: A New Understanding of Canine Origin, Behaviour and Evolution*, Prentice Hall & IBD (2001)

Patricia McConnell, *For the Love of a Dog – Understanding Emotion in You and Your Best Friend*, Ballantine Books (2007)

Joseph O'Connor and Ian McDermott, *Principles of NLP*, Thorsons (1996)

Jaak Panksepp, *Affective Neuroscience*, OUP USA (2004)

Karen Pryor, *Reaching the Animal Mind*, Scribner Book Company (2009)

Laura J. Sanborn, M.S., 'Long-Term Health Risks and Benefits Associated with Spay/Neuter in Dogs' (research paper published at www.naiaonline.org, 2007)

Sue Sternberg, *Successful Dog Adoptions*, John Wiley & Sons (2003)

Fabulous websites

Association of Pet Dog Trainers (UK)

www.apdt.co.uk

The Association of Pet Dog Trainers is an organisation designed to promote the highest standards of positive dog training in the UK. All members are assessed according to a strict Code of Practice. Find a local trainer via their regional map, or be inspired to join in with a dog sport, such as Rally.

Association of Pet Behaviour Counsellors

www.apbc.org.uk

Full members are behaviour specialists who work on veterinary referral. Visit the website for details of your nearest member.

Clever Dog Company
www.cleverdogcompany.com
Sarah Whitehead's website, featuring free downloads on behaviour and training for your dog, as well as video clips and practical suggestions for a harmonious relationship with your dog.

Think Dog!
www.thinkdog.org
Accredited education courses in canine behaviour and training. Start a new career or fuel your passion with a home study or attendance course.

Train Your Dog Online
www.trainyourdogonline.com
Train your dog with video lessons in the comfort of your own home.

Dog Star Daily
www.dogstardaily.com
Fab dog blog featuring behaviour and training experts from around the globe.

Clickertraining.com

www.clickertraining.com

Karen Pryor's excellent website focusing on all things clicker training – watch a crab being trained to ring a bell with its claw!

Learn dog language

www.dogsavvi.com

Do you speak dog? An online educational course designed to help you learn the subtleties of canine body language and facial expression.

Dr Sophia Yin

http://drsophiayin.com

Veterinary surgeon and behaviourist Dr Yin's magical website looking at the dominance controversy and domestic dog behaviour.

Sue Sternberg

http://www.suesternberg.com

Excellent website devoted to shelter dogs, assessments and training.

The Gift of Fear
www.gavindebecker.com
Gavin de Becker's books, thoughts on media-speak and articles.

Kikopup – on YouTube
www.youtube.com/results?search_type=search_users&search_query=kikopup&uni=1
Inspirational clicker trainer Emily Larlam demos amazing tricks with her own dogs.

Scentwork
http://talkingdogsscentwork.co.uk
Get your dog's nose working in the right direction with structured scent work.